Energy, Climate and the Environment

Series Editor: **David Elliott**, Emeritus Professor of Technology Policy, Open University, UK

Titles include:

Manuela Achilles and Dana Elzey (*editors*)
ENVIRONMENTAL SUSTAINABILITY IN TRANSATLANTIC PERSPECTIVE
A Multidisciplinary Approach

Luca Anceschi and Jonathan Symons (*editors*)
ENERGY SECURITY IN THE ERA OF CLIMATE CHANGE
The Asia-Pacific Experience

Philip Andrews-Speed
THE GOVERNANCE OF ENERGY IN CHINA
Implications for Future Sustainability

Ian Bailey and Hugh Compston (*editors*)
FEELING THE HEAT
The Politics of Climate Policy in Rapidly Industrializing Countries

Gawdat Bahgat
ALTERNATIVE ENERGY IN THE MIDDLE EAST

Mehmet Efe Biresselioglu
EUROPEAN ENERGY SECURITY
Turkey's Future Role and Impact

Beth Edmondson and Stuart Levy
CLIMATE CHANGE AND ORDER
The End of Prosperity and Democracy

David Elliott (*editor*)
NUCLEAR OR NOT?
Does Nuclear Power Have a Place in a Sustainable Future?

David Elliott (*editor*)
SUSTAINABLE ENERGY
Opportunities and Limitations

Huong Ha and Tek Nath Dhakal (*editors*)
GOVERNANCE APPROACHES FOR MITIGATION AND ADAPTATION TO CLIMATE CHANGE IN ASIA

Horace Herring and Steve Sorrell (*editors*)
ENERGY EFFICIENCY AND SUSTAINABLE CONSUMPTION
The Rebound Effect

Horace Herring (*editor*)
LIVING IN A LOW-CARBON SOCIETY IN 2050

Matti Kojo and Tapio Litmanen (*editors*)
THE RENEWAL OF NUCLEAR POWER IN FINLAND

Antonio Marquina (*editor*)
GLOBAL WARMING AND CLIMATE CHANGE
Prospects and Policies in Asia and Europe

Catherine Mitchell, Jim Watson and Jessica Whiting (editors)
NEW CHALLENGES IN ENERGY SECURITY
The UK in a Multipolar World

Catherine Mitchell
THE POLITICAL ECONOMY OF SUSTAINABLE ENERGY

Ivan Scrase and Gordon MacKerron (*editors*)
ENERGY FOR THE FUTURE
A New Agenda

Gill Seyfang
SUSTAINABLE CONSUMPTION, COMMUNITY ACTION AND THE NEW ECONOMICS
Seeds of Change

Benjamin K. Sovacool
ENERGY & ETHICS
Justice and the Global Energy Challenge

Joseph Szarka
WIND POWER IN EUROPE
Politics, Business and Society

Joseph Szarka, Richard Cowell, Geraint Ellis, Peter A. Strachan and Charles Warren (*editors*)
LEARNING FROM WIND POWER
Governance, Societal and Policy Perspectives on Sustainable Energy

David Toke
ECOLOGICAL MODERNISATION AND RENEWABLE ENERGY

Thijs Van de Graaf
THE POLITICS AND INSTITUTIONS OF GLOBAL ENERGY GOVERNANCE

Xu Yi-chong (*editor*)
NUCLEAR ENERGY DEVELOPMENT IN ASIA
Problems and Prospects

Xu Yi-chong
THE POLITICS OF NUCLEAR ENERGY IN CHINA

**Energy, Climate and the Environment
Series Standing Order ISBN 978–0–230–00800–7 (hb)
978–0–230–22150–5 (pb)**
(*outside North America only*)

You can receive future titles in this series as they are published by placing a standing order. Please contact your bookseller or, in case of difficulty, write to us at the address below with your name and address, the title of the series and the ISBNs quoted above.

Customer Services Department, Macmillan Distribution Ltd, Houndmills, Basingstoke, Hampshire RG21 6XS, England

Climate Change and Order
The End of Prosperity and Democracy

Beth Edmondson
Senior Lecturer, Monash University, Australia

Stuart Levy
Director, Diploma of Tertiary Studies, Monash University, Australia

© Beth Edmondson and Stuart Levy 2013

All rights reserved. No reproduction, copy or transmission of this publication may be made without written permission.

No portion of this publication may be reproduced, copied or transmitted save with written permission or in accordance with the provisions of the Copyright, Designs and Patents Act 1988, or under the terms of any licence permitting limited copying issued by the Copyright Licensing Agency, Saffron House, 6–10 Kirby Street, London EC1N 8TS.

Any person who does any unauthorized act in relation to this publication may be liable to criminal prosecution and civil claims for damages.

The authors have asserted their rights to be identified as the authors of this work in accordance with the Copyright, Designs and Patents Act 1988.

First published 2013 by
PALGRAVE MACMILLAN

Palgrave Macmillan in the UK is an imprint of Macmillan Publishers Limited, registered in England, company number 785998, of Houndmills, Basingstoke, Hampshire RG21 6XS.

Palgrave Macmillan in the US is a division of St Martin's Press LLC, 175 Fifth Avenue, New York, NY 10010.

Palgrave Macmillan is the global academic imprint of the above companies and has companies and representatives throughout the world.

Palgrave® and Macmillan® are registered trademarks in the United States, the United Kingdom, Europe and other countries.

ISBN 978–1–137–35124–1

This book is printed on paper suitable for recycling and made from fully managed and sustained forest sources. Logging, pulping and manufacturing processes are expected to conform to the environmental regulations of the country of origin.

A catalogue record for this book is available from the British Library.

A catalog record for this book is available from the Library of Congress.

Contents

List of Figures and Tables vi
Series Editor's Preface vii
Preface ix
Acknowledgements xi

Introduction: Why We Wrote This Book 1
1 Waiting for What? 14
2 Limits to Global Consensus 33
3 Governing Nature and Global Governance 49
4 A Rowdy and Unruly Community 66
5 Water, Disorder and Disrupted Development 80
6 Energy, Progress and Population 96
7 Energy and the Security Dilemma 113
8 Water, Food and Fire 134
9 Solutions, Ideas and Institutions 155
10 Rights, Responsibilities and Sovereignty 172
11 Identity, Ethics, Security and Order 189
12 Global Guardians 205
Conclusion: Why Global Responses Take Time 218

Bibliography 227
Index 249

Figures and Tables

Figures

2.1	Sites of diversity among states	45
3.1	Sectors of influence over states' international policy choices	59
3.2	Key drivers of states' goals	63
8.1	How changes in water challenge human communities	136
8.2	A changing rainfall system	139
11.1	International ethics, reason and identity among states	200

Tables

1.1	How climate changes challenge human societies	23
3.1	The primary decisions of states	56
3.2	An overview of the political debates and challenges of global climate change	62
4.1	Selected examples of new voices	72
4.2	How new voices challenge assumptions about states	75
8.1	Climate change impacts upon water resources	144

Series Editor's Preface

Energy, Climate and the Environment

Concerns about the potential environmental, social and economic impacts of climate change have led to a major international debate over what could and should be done to reduce emissions of greenhouse gases. There is still a scientific debate over the likely *scale* of climate change and the complex interactions between human activities and climate systems; but global average temperatures have risen, and the cause is almost certainly the observed build-up of atmospheric greenhouse gases.

Whatever we now do, there will have to be a lot of social and economic adaptation to climate change – preparing for increased flooding and other climate-related problems. However, the more fundamental response is to try to reduce or avoid the human activities that are causing climate change. That means, primarily, trying to reduce or eliminate emission of greenhouse gases from the combustion of fossil fuels. Given that around 80 per cent of the energy used in the world at present comes from these sources, this will be a major technological, economic and political undertaking. It will involve reducing demand for energy (via lifestyle choice changes – and policies enabling such choices to be made), producing and using whatever energy we still need more efficiently (getting more from less) and supplying the reduced amount of energy from non-fossil sources (basically switching over to renewables and/or nuclear power).

Each of these options opens up a range of social, economic and environmental issues. Industrial society and modern consumer cultures have been based on the ever-expanding use of fossil fuels, so the changes required will inevitably be challenging. Perhaps equally inevitable are disagreements and conflicts over the merits and demerits of the various options and in relation to strategies and policies for pursuing them. These conflicts and associated debates sometimes concern technical issues, but there are usually also underlying political and ideological commitments and agendas which shape, or at least colour, the ostensibly technical debates. In particular, at times, technical assertions can be used to buttress specific policy frameworks in ways which subsequently prove to be flawed.

The aim of this series is to provide texts that lay out the technical, environmental and political issues relating to the various proposed policies for responding to climate change. The focus is not primarily on the science of climate change, or on the technological detail, although there will be accounts of the state of the art, to aid assessment of the viability of the various options. However, the main focus is the policy conflicts over which strategy to pursue. The series adopts a critical approach and attempts to identify flaws in emerging policies, propositions and assertions. In particular, it seeks to illuminate counter-intuitive assessments, conclusions and new perspectives. The aim is not simply to map the debates but to explore their structure, their underlying assumptions and their limitations. Texts are incisive and authoritative sources of critical analysis and commentary, indicating clearly the divergent views that have emerged as well as identifying the shortcomings of these views. However, the books do not simply provide an overview, they also offer policy prescriptions.

The present text certainly meets all those aims, offering a very wide-ranging critique of international policy responses to climate change. It is a brave and heartfelt analysis of a key issue of our time, perhaps the key issue. Its basic question is: Are our international institutions and agencies up to dealing with such urgent and large transboundary issues? To which the response seems to be, no, they have failed on many fronts so far. The challenge has been too tough, given the complexity and interactivity of the system that needs to be changed. After all, what is required is a new global political framework, indeed a new world economic order. But, despite the massive problems, the authors nevertheless remain confident that progress can be made, given the necessary leadership, and they provide insights that may help stimulate the process of change.

Dave Elliott

Preface

We have written this book to increase understanding of why it takes so long for governments and others to agree on how to respond to the challenges of global climate change, and why it is important for them to continue to try to do so.

This book does not catalogue the environmental disasters that are the unattractive and disturbing by-products of modern lifestyles, popular notions of progress and visions of prosperity. Neither does this book detail the intricate processes of creating policies that respond to the scientific calls for action in order to safeguard the natural world. Such endeavours tend, necessarily, to be complex and impenetrable to all but those who are closely engaged in such matters. Instead, this book examines why it is so difficult for the international community to respond to global climate change. In doing so, we have endeavoured to analyse and explain some of the strategies that might ultimately provide the foundations for appropriate responses.

Responding to climate change is not simply a matter of finding cleaner, newer technologies or drafting policies and laws within states to curb pollution and carbon production. It is not a matter of protecting species and habitats or remembering to turn off the lights and put out the recycling. Global climate change requires appreciating how our rich and complex ways of life are interconnected with the natural environment and its ecosystems. It requires all of us, and our political leaders, to recognise that 'a picture of indefinite expansion is an impossible story of the future', and the well-being of future generations relies upon our willingness to compromise among ourselves to achieve sustainable and morally equitable forms of progress and ways of life (Stern, 2009, p. 10). Addressing the impacts of global climate change requires more than an appreciation of the sciences and more than carbon taxes or subsidies for new cleaner energy producers. Appropriate responses require an appreciation of how politics and political decision-making, in the pursuit of greater levels of international order, progress, prosperity and democracy, have contributed towards global climate change. This is useful because it also helps to explain why we also find ourselves poorly equipped to accept the grim consequences. While this does not, of itself, present immediate solutions, we have written this book because we believe that politics will ultimately furnish us with them. However, this possibility depends upon effective and courageous political leadership. Without it, politics will stand in the way of finding and enacting timely solutions.

We hope that the conversations begun by readers will contribute to new political solutions to these potentially overwhelming problems.

Beth Edmondson and Stuart Levy

Acknowledgements

This book is dedicated to our partners and children, who daily renew our confidence in human ingenuity, the power of conversations and will to negotiate win-win solutions to apparently impossible problems.

We hope that it contributes to conversations towards preserving secure futures for our children, their partners and their children, both already born and not yet imagined.

To Leanne, Mark, Cat, Rebecca, Ben, Peter, Alec, Aidan, Ashlyn, Lachie, Kye, Maya and Bryn.

We are also deeply grateful to our colleague Dr Rebecca Strating who generously provided ideas and suggestions concerning the scope and contents of this book, and provided us both with truly sustaining conversations, good humour and the fullest of collegial support at all times. We are especially grateful for being challenged to think more fully about the relationship between sovereignty and societies. We cannot imagine a finer colleague.

We are also grateful to Tim Flannery, author of *The Weather Makers: The History and Future Impact of Climate Change* and Australian of the Year 2007, who encouraged our endeavours by describing it as '[a]n important book on an important subject'.

Finally, we are grateful to Monash University for providing us with the felicitous happenstance of working together for the best part of 20 years and for providing each of us with a period of sabbatical, without which this book would have taken even longer to write.

Introduction: Why We Wrote This Book

This book, like many discussions concerning climate change, began with a conversation between someone with a good grasp of the history and debates about global warming and someone who had more recently begun to pay attention to them. The science was not discussed, nor debates about the validity of that science. Neither of us saw much point in second-guessing the qualified experts on the basis of secondary analysis. Rather, we became preoccupied with the question: Why is it so hard to develop meaningful strategies for addressing global climate change? If the vast majority of the world's experts agree that climate change is real, and that human activities are either contributing or driving influences, then why has climate change not been embraced as the political challenge of this century? The problem then became how to consider who should be expected to do what, and why they should do so.

As we sketched the international community on napkins, writing pads and whiteboards, adopting political, economic, technological, ethical and historical perspectives, we became increasingly convinced that climate change is the central challenge of this century and this was repeatedly confirmed through subsequent research. Addressing global climate change requires the single largest cooperative endeavour in history, but even with it, good lasting outcomes are not guaranteed. Failing to achieve cooperation and effective political leadership in extending mitigation and adaptation strategies will condemn our children, grandchildren and subsequent generations to a life that lacks the rich variety, opportunities and promises of our own comfortable and satisfying lives as members of an affluent liberal democracy. In the absence of such leadership, cooperation will achieve little as changing interests create new imperatives or new uncertainties change agreed goals.

Accepting that climate change is occurring is the easy part. It simply requires trust in expertise of the same variety demonstrated whenever we have our car serviced, make an appointment to visit a health-care specialist, send our children to school or confirm our salaries have been deposited into our bank accounts. Modern life is composed of a series of acts of trust in the expertise of those around us. Many of these acts require us to trust in forms of expertise that lie outside our own knowledge or experience. As members of a modern, functioning society, we routinely engage in a variety of acts of trust, some of which are carefully considered, while others have become so commonplace that we rarely notice them as instances of trust-giving. As observed by Shapin (1994, p. 10), '[a]ny society in which people could not routinely trust...would fall apart. It would be incapable of the complex coordination necessary for great undertakings.' We puzzled over why some people do not accept the experts when it comes to climate change and wondered whether it is because they are the bearers of grim news and we all fear they may be right.

Our conversations became points for a journal article and then chapter plans as the scale of the endeavour took shape. Addressing climate change is not simply a matter of finding cleaner, newer technologies or drafting policies and laws within states to curb pollution and carbon production. It is not a matter of protecting species and habitats or remembering to turn off the lights and put out the recycling. Addressing global climate change requires appreciating how our rich and complex ways of life are interconnected with the natural environment and its ecosystems. It requires us, and our political leaders, to recognise that 'a picture of indefinite expansion is an impossible story of the future' and the well-being of future generations relies upon our willingness to compromise among ourselves to achieve sustainable and morally equitable forms of progress and ways of life (Stern, 2009, p. 10). Addressing the impacts of global climate change requires more than an appreciation of the sciences, and more than carbon taxes or subsidies for new cleaner energy producers. Addressing global climate change requires appreciation that politics and political decision-making have created this situation and we now find ourselves poorly equipped to accept the grim news of climate change. While this thought does not, of itself, present immediate solutions to problems presented by global climate change, we have written this book because we believe that politics will furnish us with solutions through effective and courageous political leadership. Without it, politics will stand in the way of quickly finding and enacting solutions.

This book provides a considered appreciation of the political challenges of global climate change. It does not catalogue the environmental impacts of modern industrial societies upon the natural environment although these unattractive and worrying by-products of modern lifestyles and popular notions of progress must be acknowledged as consequences of how we have chosen to live. This book does not seek to detail the processes and dispositions of creating policies that respond to the scientific calls for action in order to safeguard the natural world. Such endeavours tend, necessarily, to be complex and impenetrable to all but those who are closely engaged in such matters. Nor is this book a manifesto, a plan, for what should be done by highlighting particular values and technologies in a rally call that privileges and flatters particular visions about how our future should unfold.

Instead, this book sets out to consider the question: Why is it so difficult for the international community to respond to global climate change? Much of the above will be touched upon, as part of the answer, but in a manner that avoids the necessary complexity expressed by those who work intensively within those fields. We believe this is possible. If not, there are dismal prospects of a consensus emerging to address global climate change.

Democracy

The range of political choices now faced by states in addressing global climate change is more complex than many advocates of alternative future visions suggest. Solutions do not lie simply in finding new technologies to replace old ones (such as cleaner sources of electricity production), or finding ways to offset or obviate the impacts of global climate change. Nor are they limited to finding alternative clean energy sources, or preventing the growth of new industrial centres in developing states. Rather, the heart of the issue lies in finding new political visions and structures to support new economic systems and production practices that are less environmentally disruptive. These visions and structures have to be created collectively, and this is problematic because they rely upon agreements between many hundreds of stakeholders. Additionally, actions need to be 'consistent with rapid development' as noted by Stern (2009, p. 3), otherwise 'we will not get the global agreement that is fundamental for success'. Tensions are inherent in the need for rapid responses that are collaborative and inclusive, and herein lies the threat to democratic participation. How will many hundreds of stakeholders bring their constituencies to a timely consensus regarding appropriate responses to global climate change to ensure equitable future development?

Global climate change presents major challenges to many of the political structures, processes and practices of the international community. It threatens democratic political visions, beliefs and practices that underpin the roles and capacities of core actors, most obviously states, as well as altering patterns of land habitability, energy production and consumption and accessible water supplies. As argued in Chapter 2, consensus-based decision-making processes and the political structures and institutions they support will not achieve 'effective, efficient' international climate change responses (Stern, 2009). The political impacts of global climate change go beyond the challenges they pose for the forms, locations and distributions of human societies and their centres of production because they also challenge core political values and ideals.

The prospect that core political values are challenged as a result of global climate change impacts is a dawning realisation that few political actors readily accept and acknowledge. While slow responses can sometimes be attributed to a lack of political will, it is more often the case that policymakers become caught in uncertainties concerning risk assessments and fluctuating prospects of effectively managing changed practices. Nobody wishes to gamble with future prospects for prosperity among national and regional populations. No state wishes to burden their political processes or citizens with provisions to address climate change that constrain their national interests. Yet addressing climate change really does mean changing the ways we act, our reasons for doing so and our senses of responsibility. These all require changes to the ways we allow our interests to be politically expressed and pursued, and how we balance our interests alongside the rights and interests of others.

Global climate change challenges current visions of societies and the beliefs they hold concerning individual and collective means, rights and prospects for order and prosperity. These political challenges are both deep and complex, rendering a great many aspects of contemporary economies, patterns of human settlement, production and overall lifestyles vulnerable to disorder. However, these challenges are not limited to matters of production, energy consumption and the health-related consequences of climate change impacts. Rather, they go to the heart of beliefs concerning the relationship between developed and developing societies, present and future societies, viable forms of human communities and the roles of natural resources in securing their longevity. Presently, various actors grapple with particular aspects of climate change and their efforts are admirable. However, many of these tend to miss the central point that addressing global climate change requires the contemporary world to relinquish particular political visions and simultaneously generate new visions to support viable futures. This is an intensely difficult project.

The key distinction between states and other actors lies in the political responsibilities that states hold, their commitments to ensure the protection of their people and territory and to promote wellbeing within their societies. These are fundamental goals for sovereign states and their abilities to maintain self-determination goals into the future rely upon these activities. While non-state actors fulfil a range of requirements in the modern integrated political community, and are often crucial in preserving orderly conduct among states, they remain dependent upon the validation of sovereign states (Hoffman, 1997). At one level, modern complex ecological interdependence can be seen to 'bind all nations...even the largest, most powerful ones...[to] the community of all other nations' (Biermann and Dingwerth, 2004, p. 6). Yet, such ecological interdependency is likely to add intensity to the competition among states as they seek to advantage themselves and further their own interests and goals. Nonetheless, in the 21st century, ecological interdependence among states represents an even more important development than collective security.

Despite the presence of new and louder voices in the international political community, sovereign states retain primacy by virtue of their decision-making authority and their material resources. States no longer display 'hard shell' boundaries and their internal national interests are no longer the sole considerations in decision-making (Edmondson and Levy, 2008). This is especially the case among developed liberal democratic states that are influenced by groups that are not cartographically prescribed. These non-state voices exist within and alongside states, and the manner in which states respond to them can influence, and be influenced by, their identities and interests. The advantages of listening to their contributions lies in the interest, expertise, passion and popular support they can galvanise. In so far as there is a disadvantage to the open democratisation of deliberations about global responses to climate change, it is the sheer volume of expressed interests that need to be accommodated. Existing international and national political processes genuinely struggle to produce timely responses from such a cacophony (Edmondson, 2009).

In the future, states that do not respond to the problems of climate change are likely to experience severe constraints upon their autonomous authority. Such a development could entail a further evolution in the rights, responsibilities and practices associated with states' sovereignty. Multilateral action, to support coordinated and collaborative climate changes responses, could override the norm of non-intervention and thus diminish the notion of absolute autonomy that states have historically enjoyed. The future actions of states will also be guided and challenged by the voices and actions of other states, intergovernmental organisations (IGOs) and non-state actors, including corporations and non-governmental organisations (NGOs).

It may become plausible for state sovereignty to evolve without rights of non-intervention as the forms and means of collectively recognising unique political identities are altered by global responses to climate change. Recognition of sovereign statehood could then be based upon respect for, and participation in, policies to ensure domestic and international protection of the global environment. Criteria for statehood already exist: for instance, admission to the United Nations is dependent upon declarations of respect for international law and notions of collective security (Brownlie, 1998; United Nations, 1945). These require states to acknowledge particular codes of conduct in order to maintain the international community and their place within it. However, the scope of global climate change, the array of impacts it poses and their organic nature threatens the natural environment that sustains this community and, within varying timeframes, the states that compose it. Consequently, preserving the integrity of the global environment and international community may become a more fundamental imperative than preserving the political features, rights and capacities of individual states. A rationale for accepting this evolution of state sovereignty may emerge from recognising that sovereign statehood is no longer an 'inviolable principle' (Snow, 2006, p. 85) in the face of more fundamental challenges.

The political ramifications of global climate change extend to the need for new understandings of the roles, rights and responsibilities attributed to states. The contemporary international political community has been formed on the basis of ideas concerning states as primary sites of political authority that, in theory, relate to each other according to democratic principles. Each state is an autonomous, authoritative member of the international community that can freely express its views and peacefully pursue its own interests, at least within the confines of its territory. While some divergence between theory and practice always exists, for the most part it is this belief that underpins international relations. For the international community to effectively address global climate change, fundamental political transformations will be necessary, and the extent of these may call into question the viability of maintaining democratic processes for international cooperation. Addressing the complexities of climate change is a global political endeavour further compounded by expert opinions that insist the timeframes for responses are shortening. At the end of the day, what emerges as effective responses will be largely determined by whose voices are heard.

Order

Maintaining international order requires international actors to work towards common visions of a desirable future. From shared understandings, norms and treaties are developed by political actors grappling with a broad array of interests and influences, and their rights to authority reflect their responsibilities for maintaining order. Greater significance is likely to be placed upon developing and adhering to global environmental norms and international laws because citizens and states increasingly perceive effective management of climate change impacts as important to their future well-being. Issue areas concerned with the environment, such as global weather patterns and water distribution, are becoming increasingly subject to the development of new international institutions, conventions and treaties. As a consequence, the legitimacy of sovereign states becomes increasingly contingent upon their contributions to reducing climate change effects. Under these circumstances, political order and security depend upon the construction and reinforcement of international norms, agreements and risk assessments concerning global climate change.

The existence of common goals defines a society and within the international community the predominant shared goal between states is order. Order can be defined as a pattern of activity that leads to a particular arrangement of social life designed to uphold the goals shared between states (Bull, 2002, pp. 3–4). Goals are important because they outline the kinds of actions that states are expected to take to be considered legitimate. They reinforce the interconnectedness of states in modern global politics and routinely lead to negotiations and acts of cooperation. Intergovernmental organisations then engage in a diverse range of activities, including gathering information, collecting data and creating shared research and monitoring programmes. This shared knowledge then becomes the foundation upon which states negotiate and seek to create regulatory mechanisms to sustain international order.

The activities of knowledge- and policy-based communities impact most significantly and directly on negotiation and bargaining in creating treaties, conventions and international organisations. Multilateral, collective action agreements, such as treaties and regimes provide sites for negotiation in which the interests of parties are revealed and refined. Negotiation and bargaining thereby provide vehicles for collective problem solving between states, and their effectiveness, in this regard, can be seen in the implementation of various agreements concerning transboundary pollution, hazardous wastes and ozone-depleting substances in the late 20th century. In other areas, such as carbon pollution reduction and greenhouse emissions, extensive negotiations have failed to produce effective collective action agreements.

Environmental refugees became accepted episodic political and humanitarian challenges during the 20th century. In the 21st century, they may well become much more common and highly visible consequences of global climate change. Dealing with the people displaced by climate change and population increases will require, or forcefully precipitate, loosening sovereignty as a primary organising principle of the international political community. Geographic borders, as symbols and practical attributes of politically organised sovereign states, may no longer be morally defensible or ethically sustainable because retaining them would see the stateless victims of climate change locked out of secure environments. This prospect raises moral and ethical challenges on a scale not previously encountered by the modern rights–derived global community.

A failure by states and the international political community to address the effects of climate change and water redistribution will see their territories, resources and societies impacted by global changes to weather and water conditions. Security of urban water supplies, rising sea levels, the effects of extreme weather events and climate effects upon global food security are now recognised as interconnected and significant issues (see Chapter 5). At the same time, the global population is rapidly increasing as the effects of global climate change become more difficult to reverse and are more widely experienced. Despite declining fertility rates in many developed and westernised societies, forecasts continue to predict a significant increase in human numbers such that the global population will likely rise by a third between 2006 and 2050, from 6.1 to 8.9 billion (Edmondson and Levy, 2008).

Future population growth will create significant problems for densely populated states. Others, such as the island states of the Pacific and those with significant areas below sea level, will struggle to contain their citizens within borders impacted by the effects of changes in sea levels. According to the fourth report of the Intergovernmental Panel on Climate Change, the 20th century saw a global rise of sea levels that contributed to 'increased coastal inundation, erosion and ecosystem losses' (Nicholls et al., 2007, p. 317). For other states, changes in climate, weather patterns and water distributions may jeopardise or disrupt the availability of sufficient food supplies raising the spectre of future famines and forced migrations within and across state borders.

Perhaps most disturbing is that over the next 50 years the largest population increases will occur among developing states where infrastructure is already deficient (United Nations Development Programme, 2008). These states will be among the least able to address the effects of global climate change and water redistributions making them likely to become sites of large-scale, climate-impelled human tragedy. It seems unlikely that these effects will be experienced exclusively through the media and confined to the television screens of the developed world. As the burden of housing and feeding large numbers of climate refugees shifts to the more developed and less immediately impacted states, social tensions are likely to increase. Nationalism, racism and lifestyle protectionism may then mount serious challenges to liberal ideals concerning universal human rights, equality and access to security.

Maintaining order within a climate-changed world will prove difficult as states pursue new means to define and proscribe jurisdictions and people, especially in overpopulated and underprivileged regions. The Intergovernmental Panel on Climate Change (2007, p. 317) notes that:

> Developing nations may have the political or societal will to protect or relocate people who live in low-lying coastal zones, but without the necessary financial and other resources/capacities, their vulnerability is much greater than that of a developed nation in an identical coastal setting.

The burden of maintaining global order in this instance could be expected to fall upon industrialised 'Western' states that have also been expected to assume greater degrees of responsibility for lowering carbon emissions (Davis and Caldeira, 2008; Roberts and Parks, 2007). For their part, developing states potentially find themselves sacrificing industrial development and economic growth in order to participate in international efforts to mitigate climate change effects. While such contributions initially seem admirable, they would inevitably create social instability and turmoil among vulnerable communities and create further problems for states with short histories of political and economic stability. Efforts to secure the climate must then be measured against the human costs of increased poverty and insecurity among the least well-off and must also recognise that without preserving international order, little benefit can be achieved.

Orderly international conditions enhance states' prospects of self-preservation and mutually beneficial relationships. By establishing preconditions for relations between states and their authoritative status, the self-perpetuating norms of international order reinforce states as primary political actors. International order subsequently arises from and establishes conditions within which states agree to compromise and reconcile competing interests to experience the benefits of a relatively stable political environment. As such, international order transcends individual security and economic interests, and it shapes the conditions under which peaceful coexistence might be preserved. International order arises from the collective efforts of states to ensure security for their citizens, protecting them from violence, invasion and government by foreigners.

An international society can be said to exist when states, conscious of their common interests and common values, conceive themselves to be bound by a common set of rules, and they share in the workings of common institutions (Bull, 2002, p. 13). The contemporary international community clearly displays these features through the legal equality of states, their rights to fixed borders and reliance upon recognition for functional international membership. Nationally, societies consist of complex sets of relations across political, economic and social spheres. Globally, the complex interdependences that define relationships between states create a society of sorts (Luard, 1990, p. 3).

The need for order is at once a goal and a requirement of an international society, and this means that norms and institutions play integral roles in the operation of international society. At the core of international society are 'principled rules, institutions and values that govern both who is a member of the society and how those members behave' (Finnemore, 1996, p. 18). These international institutions can be defined as 'persistent sets of rules...that prescribe behaviour roles, constrain activity and share expectations' among states (Keohane, cited in Reus-Smit, 1997, p. 557). Similarly, norms are practised, habitual undertakings that 'regulate conduct, by setting up regularities in behaviour' (Luard, 1990, p. 62). These norms and institutions are politically constructed mechanisms, through which states cultivate predictability and stability in their relations. These sustain patterns of activity that constitute order and socialise states to construct and accept rules of practice.

Progress and prosperity

Global climate change exposes the myth of progress and human control over our environments. Its political ramifications challenge the foundations of modern human societies and their ideas about what constitute ethical and sustainable notions of progress and prosperity. Scientific knowledge has revealed links between the processes by which industrial progress has been achieved and serious and extensive damage to the biosphere that now threatens massive disruptions to modern ways of life (United Nations Development Programme, 2008; Parry et al., 2007). More than this, however, changes to the planet's climate affect the distributions and activities of human societies. Whether human activity drives climate change or not the truth remains that '[a] transformation of the physical geography of the world also changes the human geography: where we live, and how we live our lives' (Stern, 2009, p. 9).

The global ecosystem, artificially divided by political borders, is reduced to a resource repository that states utilise in expanding profits, consolidating security or enhancing their influence or capacities in relation to others. However, the consequences of global climate change reveal the vulnerability of the earth's ecosystems and the reliance of humans upon them (Anderson and Leal, 1998; Brown, 2005). Human societies, their economic endeavours and political systems are thoroughly reliant upon nature, including the specific features of their physical environments. Most importantly, the nature of states as sovereign authorities poses additional obstacles to the attainment of effective climate change responses.

The capacities of states have, to a certain extent, been historically founded upon their natural endowment of resources. More recently, technology and political affiliations arising from the development of a global international economy allowed states to defeat the tyranny of distance and access resources and markets beyond their own borders. The current international community, however, now faces threats to these practices from the long-term impacts of climate change. The reallocations of natural resources across states' political boundaries, the relocation of cost-effective trade routes and new weather patterns that disrupt the supply of natural resources are obvious effects. Energy-hungry forms of transport and economic activity that contribute to large volumes of greenhouse gases will, over time, attract economic disincentives that significantly restructure international economic activity. These developments will affect the capacities of states to maintain domestic standards of living and to project power and influence abroad.

The costs of climate change (managing the effects of natural disasters and devising environmentally responsible domestic development strategies) will increasingly absorb the energies and resources of states. Many states' capacities for foreign engagement may diminish as they look inward for solutions to address localised problems. Catastrophic effects of severe weather events, such as prolonged drought, firestorms, tropical storms and flooding diminish the GDP of states and divert their resources to the domestic provision of humanitarian relief. Only the largest and/or most developed states will be in a position to simultaneously deal with the effects of climate change while retaining the resources necessary to engage with the broader international community. In such a world, mechanisms for collective action will be at a premium. At the same time, it is likely that states will increasingly seek to exert their sovereignty over the resources they have or to which they can lay claim.

Conclusion

In spite of these complexities and the challenges of balancing rights to freedom and prosperity alongside the need for international order, we believe that solutions will be found to the political impediments associated with global climate change. We remain confident that the pressing imperatives for climate change mitigation and adaptation policies will create a political 'tipping point' that will lead governments and other international actors to new political solutions. Effective and durable policies need to pay attention to the sensibilities and characteristics of the diverse actors and stakeholders whose views must be incorporated into proposed solutions. Those will need to accommodate the diverse rights, responsibilities and perceptions of common good that exist and are shared within the international political community (Najam, 2005a, p. 245; Bonanate, 1995). However, physical interdependencies do not always make it easier to achieve international agreements as they present additional questions concerning who has which rights and responsibilities, and who holds authority to decide upon their distribution (Young, 2002).

Achieving effective international responses to global climate change relies upon a morally defensible and broadly accepted distribution of the costs to development and modern consumption-based lifestyles. These require states to identify broadly accepted frameworks of political association and authority. They also require collective actions by states and international institutions to achieve internationally coordinated procedures to harness the activities of numerous and diverse actors. Their success depends upon accepted instruments and processes to compel states and other actors to comply with agreed targets and other initiatives arising from international agreements. Efforts to develop mitigation and adaptation policies to alleviate climate change impacts thus represent significant political challenges for the international community. Effective, efficient and equitable climate change responses require new beliefs concerning the roles of states to ensure that they accept new responsibilities for the well-being of other citizens and new authority over other social and economic actors.

1
Waiting for What?

Introduction

Through much of 2009, the world waited for political leaders to convene at a climate summit in Copenhagen. In turn, many of these leaders waited for others to lead the way in declaring emissions targets and other political strategies for responding to climate change. Overall, 2009 was a year in which the world waited for a new program of action to emerge. Some held high expectations of likely comprehensive agreements. Some expected that the Copenhagen Summit would provide a venue for intense political bargaining. Many believed that a raft of linked agreements would emerge from what was expected to be one of the most important meetings of government leaders since the first international climate change discussions of the 1980s. Even those with relatively modest expectations anticipated the likely negotiation of parameters for a new global climate change agenda. Few expected that this cornerstone event in 21st-century political leadership would result in so little. The Copenhagen Climate Summit provided no substantive new emissions targets, no new international standards or strategies and no new political alliances or coalitions of interests to support subsequent initiatives.

Instead, the Copenhagen Summit produced an agreement to wait-and-see how climate change consequences unfold, how and which states meet their previously set targets and how the weight of scientific consensus reconfigures projections of likely climate change impacts. In effect, this amounts to a general political preference for doing nothing to respond to climate change in order to protect a business-as-usual pursuit of economic prosperity. The failure of government leaders to commit their own peoples and industries to reducing greenhouse gas emissions can only be understood in terms of wilful non-response. The wait-and-see approach adopted at Copenhagen was chosen by political leaders of societies who prefer to continue to enjoy the comforts and prosperity afforded by immediate and ongoing mass consumption of fossil fuels. Their non-response positions derived from clinging to hopes of easy solutions or the possibility that costs will be borne by others because these political leaders deemed the short-term political costs of pursuing meaningful climate change adaptation and mitigation responses too high. At present, intensive industrial production and the maintenance of an international political economy are privileged over other imperatives because progress, prosperity and democratic freedoms override other considerations. Consequently, global climate change will continue to impact upon the earth, its populations and forms of human societies for decades to come with escalating impacts upon habitats and livelihoods.

Analyses of the Copenhagen Climate Summit as short-sighted shifting of hopes towards new technologies, or a lack of willingness among affluent states to accept large-scale changes to their economies and lifestyles, miss the central point. These failures are derived from an unwillingness to take unilateral actions and assume leadership roles as well as an absence of shared alternative political ideals and visions across the world. It is not merely in the context of global climate change that leaders of governments have revealed preferences to let others bear costly burdens. The histories of international peacekeeping, pollution management and uneven economic development are replete with these political dynamics.

Effective climate change responses are only likely to emerge following the exertion of pressure upon political and economic actors to demonstrate new forms of political leadership that seek to preserve human societies into the future. Contemporary states are particularly sensitive to political pressure from other states and intergovernmental organisations, and these might serve to support new expressions of political leadership. At present the diversity of states creates obstacles to the collective decision-making that was previously anticipated in international climate change conferences and forums. Diversity increases states' tendencies to cling to historically based political visions and relationships as sources of effective order to support their continued territorial independence. These political realities are not new, so closer observation of the events leading up to the Copenhagen Climate summit should have enabled very limited outcomes to have been widely anticipated.

This does not mean, however, that we should accept non-responses as either inevitable or tolerable.

The leaders of modern sovereign states are poorly equipped to demonstrate the forms of political imagination that are now required for responding to global climate change. The mainstays of international political dealings between states, such as trade, diplomacy and collective security, are poor vehicles for political change, although they have proven relatively effective at enabling predictability in relations between states. This quest for international stability contributes a powerful inertia in international settings where changed political values and visions are involved. Stability and predictability constitute powerful international dynamics because they support economic well-being and security. Many states are currently incapable of de-coupling their hopes of economic prosperity through industrial production from beliefs in the independence of statehood. These organising principles and patterns of conduct between states are unlikely to support changed forms of relations and visions of well-being.

Climate change, the availability and possession of useful resources and maintaining global order present particular challenges for world peace. Avoiding the worst consequences of climate change and redistributions in resources will be improved by revisiting the international political community's ideas, values and sources of political authority and by re-examining states' power and capacities to provide security. World responses to global climate change, and the major economic and political reconfigurations arising from it, will be determined by the sources and dynamics of major tensions between states and their opportunities and abilities to promote the peaceful resolution of conflict. Climate change presents significant challenges for the relations between states, including their orderly conduct in recognising other sovereign authorities and their continued general adherence to principles of non-intervention (Edmondson, 2011; Barry and Eckersley, 2005; Postiglione, 2001; Mathews, 1991).

Changes in the patterns, volumes and distribution of rainfall and corresponding increases in droughts and floods will impact upon the availability and locations of arable land and useful water resources (Dow and Downing, 2007). These changes will alter where people might live, where and how cities might be sustained and how the world's populations might support themselves, including through food production (Romero-Lankao, 2008; Roaf et al., 2005). Additionally, warming oceans affect marine-life stocks, their breeding and food cycles, ocean currents and their cooling effects. These changes shape subsequent weather patterns because the oceans are key drivers of wind, rain, cloud formation and the earth's abilities to absorb carbon dioxide.

Rising sea levels will ultimately impact upon all land masses, and this basic change in the physical world will have enormous political ramifications. At present, political leaders tend to treat rising sea levels as limited events that present challenges for particular states. The most notably affected are low-lying states that are surrounded by water, or those with especially wide river mouths where ocean inflow presents risks of freshwater contamination and flood (Dow and Downing, 2007). However, rising sea levels will challenge all states, and not just because of human migrations.

Rising sea levels are linked with the many other aspects of global climate change, most importantly, with changes to the geophysical drivers of climatic patterns that are most prone to high levels of unpredictability. These present the most difficult challenges for developing mitigation and adaptation strategies. Such changes, even if experienced at modest levels, disrupt orderly relations and structures within the international political community. The effects of climate changes will impact upon the circumstances and experiences of states and alter their traditional capacities.

Global climate change is presenting the international political community with a need for fundamentally different approaches in the ways that states seek security. It presents increasing challenges to states' abilities to secure their identities and achieve their political and economic interests. These challenges also risk undermining states' contributions to the ongoing maturation of the international political community. It raises further questions as to the roles that might be played by those states that, for reasons of ideology or capacity, are unable to make meaningful contributions to environmental issues which are becoming pressing. Although some states may be in a position to undertake leadership roles, it seems inevitable that others will lack the political adaptability required to lead effectively in this new environment. Many will struggle to demonstrate timely responsiveness to information, and some will resist attempts to pursue responses under the leadership of others.

Large-scale resource and territory wars are not inevitable in the 21st century, but neither is their avoidance assured. Minimising conflicts arising from fresh water and energy shortages will rely upon political management of social and economic vulnerabilities (see Chapters 5–7 for further discussion of these issues). Maintaining international order relies upon the international political community becoming a guardian of the global ecosystem through governance mechanisms that overturn accepted divisions between domestic (internal) and international (external) activities among states. Central to this are the questions: At what point do states and the international political community accept responsibility for initiatives to address climate change and its consequences? When will they exercise the political leadership required to achieve mitigation and adaptation strategies to preserve human communities into the future?

The challenges of increasing atmospheric and ocean temperatures, more frequent and more severe storms, shortages in useful water supplies and problems of maintaining orderly relations among states require the reinvention of international structures and accepted modes of conduct among states (Dyer, 1996). Global climate change presents the international political community with major challenges that further reduce the utility of pre-existing understandings of order and the nature of sovereign states (Gardiner and Hartzell-Nichols, 2012, pp. 1–3; Diehl, 1997). This is only partly because climate change consequences transcend borders and potentially disorder our understandings of security. Maintaining fixed borders, protecting people and resources can no longer be held as central roles of states when they can no longer be secured or come to be directly threatened by environmental challenges. If mitigation and adaption strategies are to reduce the extent and magnitude of global climate change consequences, they will rely upon decision-making capacities of states that overcome the dilemmas arising from representative political systems, modern economic production and consumption. Solutions will rely upon international agreements that recognise unequal capacities for adaptation among states and accommodate notions of justice (Markowitz and Shariff, 2012a; Page, 2006). Leaving some states behind will not provide good outcomes for any.

As the effects of climate change become more pronounced, states and the human communities they contain will experience new costs and limitations arising from their geographic locations (United Nations Development Programme, 2008). Agricultural practices and viability will be jeopardised in some regions of the world as weather patterns change and the amount of land suited to food production is reduced. The need to curb reliance upon fossil fuel use, in industry and transportation to minimise further greenhouse gas emissions, will also impact upon what can be produced, where and for what cost. The increasing incidence of extreme weather events, such as more intense tropical cyclones and storms, are projected to affect larger areas as they move further toward the poles. Similarly, heatwaves, drought and flooding from storms and concentrated rainfalls are predicted to become more commonplace (Intergovernmental Panel on Climate Change, 2007b; Gore, 2006; Flannery, 2005).

All such events will have economic and social costs (United Nations Development Programme, 2008; Stern, 2006). Slowly rising sea levels will inundate low-lying coastal regions and urban centres. These will disrupt daily life and national economies as agriculture, industries and human settlements relocate. Changes such as these will impact the global population in countless ways and force political responses as demands grow for policies and actions to address their consequences and underlying contributing causes (United Nations Development Programme, 2008).

The political composition of states, and the international political community constructed by them, emphasise states' rights to independent decision-making and have nurtured hopes of economic prosperity through industrial production and trade. The ideas that underpinned states' behaviour and identity have created an international political community that pursues order and security through economic expansion (Philpott, 2001). Chief among these ideas has been respect for the sovereign integrity of states, which was understood to depend upon widespread acceptance of states' rights to be protected from intervention by others, especially in their internal affairs (Najam, 2005a; Kütting, 2000; Hoffman, 1997; Morgenthau, 1973). This right underpinned their pursuit of industrial development and economic progress and has constructed an international political community with limited capacities for responding to global climate change. The contemporary global political issues of addressing climate change and maintaining order reveal the possible end of autonomous security, upon which sovereign statehood relies.

For modern states, everything – the provision of food, shelter and employment for the population, together with production and the balance of trade – depends upon the availability and management of an extensive list of resources. Sovereignty, security and economic growth are all intimately aligned with resource use. Concerns about long-term land and resource use are largely a relatively recent addition to the list of state responsibilities. In the face of 21st-century climate change and resource redistribution, scarcity and competition become matters over which even the most powerful states can exercise only limited control.

The nature of the international political community partially explains why it continues to struggle to achieve effective international agreements for global climate change mitigation and adaptation strategies.

The manner in which collective responses to problems are determined and enacted has been premised upon expectations that major threats to well-being will affect only groups of states. In the 21st century, these expectations are affecting economic, social, cultural and strategic relations between states and these impacts will increase as global climate change impacts unfold. In spite of some obvious links between environmental management and governance, many states have shown themselves reluctant to accept responsibility for the environment except in relation to resource use. As noted by Page (2006, p. 4), 'the popular view has been that the bad effects of weather are regrettable but not unethical, inequitable or unjust...since they are no one's fault'. This confirms attitudes towards the physical environment as one of the 'givens' within which governments conduct themselves and pursue their visions of economic prosperity and social progress. In most societies, the environment provided a backdrop to political affairs, constituting a source of valuable resources and a recreational facility (Ponting, 1991).

Important political structures

Old concepts and systemic dynamics no longer ensure orderly conduct in international relations, and, as the diverse consequences of global climate change highlight inequalities, we might anticipate the emergence of new political divisions and attempts to establish new structures. New major powers are likely to be those that accommodate and positively respond to these new challenges. New understandings of international relationships, actors and processes will be important for 21st-century political practices as pre-existing dynamics, such as beliefs in progress as a measure of effective government and security through non-intervention lose utility. Likewise, current security preoccupations with religious and ethnic identities as sources of tension between states, and likely security risks within them, offer little that is useful for appreciating how the international political community should respond to contemporary global climate change and related environmental concerns. In the 21st century, as climate change and water resource issues impact more directly upon states' capacities, it will be important to understand states' abilities to generate new modes of production, forms of political authority and accepted behaviour in cooperative decision-making.

Autonomous security, articulated in the form of independent sovereign statehood, was previously considered a rational means of securing political independence (Waltz, 1979). Over time it produced an international political system that contributed both directly and indirectly to environmental destruction and created an array of structural problems. Sovereignty created ideas of independence, economic progress and security that produced actions among states that were intelligible and defensible during the 19th and 20th centuries. Among these were ideas about resource security and exploitation, progress as a competitive measure of comparative civilisation and the pursuit of rival policies to promote physical security and social identity. In the 21st century, however, many of the long-term consequences of these ideas and actions have become dangerous and threaten the future viability of states, the international political community and the global environment.

The expanding size and range of activities undertaken by human communities in pursuit of higher standards of living constitutes 'a potential threat to the environment' (Jackson and Sorensen, 2007, p. 256). Human action supported by sovereign states has resulted in 'widespread ecological degradation' (The Commission on Global Governance, 1995, p. 29). The natural environment is no longer entirely natural; it has been shaped and impinged upon by successive human communities that have changed its composition and functioning. Whether or not human communities have been wholly responsible for the onset of climate change, or have merely contributed to a range of causal factors, is beside the point. The consequences of changes in the global ecosystem will have to be dealt with by international political actors, including states, engaged in communal problem solving.

Dividing the international political community into sovereign territorial states and an expanding host of 'other' actors with lesser rights, obligations and prerogatives may prove detrimental to timely climate change responses. While states remain central actors, they are also problematic sources of authority because their independence and drives to protect their own people, territory and resources ahead of the needs of others creates enduring divisions and limits communal problem-solving (Gardiner and Hartzell-Nichols, 2012; Edmondson, 2011). These dynamics and their impacts upon international structures and concepts cannot be ignored because they arouse passions within states and among international actors, reinforcing the importance they attribute to their resource security rights rather than establishing structures that promote shared responsibilities. The environmental, social, economic and political consequences of climate change require the re-evaluation of dominant conceptions about the nature of progress and desirability of democracy because these views entrench resource use and production practices determined by assertions of rights with little regard for costs to others.

Debates about who holds responsibility for environmental problems arising from climate change are internationally divisive. Developed states are responsible for modern economic development practices that consume natural resources to maintain high standards of living and place increasing burdens on the environment. Developed states produce disproportionate amounts of waste and their industrial and agricultural practices pollute and degrade the natural environment (Gardiner and Hartzell-Nichols, 2012; Roberts and Parks, 2007; Flannery, 2005). Developing states also contribute to climate change through their high rates of population growth and unregulated pursuit of industrialisation. The pressing effects of climate change do not allow time for such allocations of blame that fail to progress appropriate responses.

The elephant in the room

The fourth report of the Intergovernmental Panel on Climate Change (IPCC), released in 2007, further advanced understandings of predicted changes in the earth's climate due to natural variability and human activity. Drawing upon large amounts of new data, new and more comprehensive modelling techniques and a more refined appreciation of variables, it suggested that changes in the earth's climate cannot be solely attributed to natural events. Increased global temperatures are unequivocal and there is a very high confidence (90 per cent chance of being correct) 'that the global average net effect of human activities since 1750 has been one of warming' (Intergovernmental Panel on Climate Change, 2007a, p. 3). The contributions from human activity stems primarily from varieties of greenhouse gases that have entered the atmosphere from the use of fossil fuels, changed patterns of land use and modern agricultural practices. These greenhouse gases have changed the manner in which the atmosphere operates with a net effect that increases in temperature are discernible at both global and continental levels.

This report noted climate changes that included a global reduction in mountain glaciers and snow cover, a 0.17 metre rise in sea levels over the 20th century, widespread changes in patterns and amounts of rainfall, with some regions receiving more and others less, longer and more intense droughts since the 1970s, less frequent cold days and nights and more frequent heatwaves (Intergovernmental Panel on Climate Change,

2007a, pp. 5–8). The likelihood of a human contribution to these trends has been estimated to be at least 50–90 per cent, and the likelihood that these trends will continue into the future is greater than 66–99 per cent. Whether or not human activity is driving these changes or merely contributing to them, the fact remains that from the 21st century onward the natural environment will be quite different.

Changes in temperature are significant for the impacts they have on the natural environment and weather patterns both currently and into the future. It is suggested that 'past and present anthropogenic carbon dioxide emissions will continue to contribute to warming and sea level rise for more than a millennium, due to the time scales required for removal of this gas from the atmosphere' (Intergovernmental Panel on Climate Change, 2007a, p. 17). As a consequence, while current effects may be difficult to discern with certainty, giving rise to passionate debate from some quarters, they will continue and compound into the future. Many features of the present international community reflect the environment of the past, from the distribution of populations to the drawing of state borders to secure resources. As the natural environment changes, so too will the units, associations, roles, obligations and expectations of the actors within the international political community (Table 1.1).

Ongoing perceptions of sovereign territorial states as independent actors, able to autonomously provide for their own security, will frustrate attempts to cultivate the necessary levels of concern and cooperation that are required. However, states are not merely forms of association with supreme judicial authority within their borders; they

Table 1.1 How climate changes challenge human societies

Challenge to production and progress	Related climate change impacts
Agricultural production and food security	Water availability, changed rainfall patterns, expanded dry-land zones, loss of arable land
Water stress and water insecurity	Changed rainfall and run-off patterns, changed river flows, changed river salinity, increased glacial melt
Rising sea levels and warmer oceans	Inundation of low-level islands and coastal areas, intense tropical storms, species depletion
Health	Heatwave deaths, spread of malaria, dengue fever and increased malnutrition

also hold responsibilities for the well-being of their citizens. It is altogether reasonable to expect them to make and enforce policies and laws that reflect their unique authoritative status. The voluntary actions undertaken by other actors within international civil society, such as corporations, NGOs and sub-national authorities, have limited capacities to impact upon emission levels at national or regional levels, although their acceptance of state-based policies is essential (Metz et al., 2007, p. 29).

Numerous studies have indicated a 'substantial economic potential for the mitigation of global GHG [greenhouse gas] emissions over the coming decades, that could offset [their] projected growth... or reduce emissions below current levels' (Metz et al., 2007, p. 11). Significant political resolve, however, will be required to achieve such outcomes. Changes in lifestyle, behaviour and cultural patterns can make important contributions to addressing climate change through emphasising resource conservation and reformed travel patterns (Metz et al., 2007, p. 17). Initiatives such as these will need to be cultivated within states for the benefit of the global community. This provides an important point of departure for traditional understandings of the political relationships between international actors. Now, as never before, members of the international political community hold very real vested interests in the internal activities of all states. What states do within their own borders and how they choose to invest and develop will be of genuine interest to the entire international political community.

Two further variables jeopardise political resolve to act on climate change issues. First, many of the effects of climate change will not be felt within the political lives of contemporary decision-makers even though they will experience the potential political backlash from the costs of mitigation policies and strategies (United Nations Development Programme, 2008). Second, 'the economically optimal timing and level of mitigation depends upon the uncertain shape and character of the assumed climate change damage cost curve' (Metz et al., 2007, p. 27). This ongoing uncertainty continues to make it difficult to convince climate change sceptics in positions of political authority that the time for action has arrived.

There is widespread agreement among the scientific community and much evidence to indicate that '[g]lobal greenhouse gas... emissions have grown since pre-industrial times, with an increase of 70 per cent between 1970 and 2004' (Intergovernmental Panel on Climate Change, 2007c, p. 3). Even with the adoption of mitigation strategies and policies to limit the future production of greenhouse gases through sustainable development practices, it is generally agreed that, from the current evidence, greenhouse gas emissions will continue to grow over the next few decades (Metz et al., 2007, p. 4). Projections for the future suggest that global warming of about 0.2°C per decade will continue over the next two decades: '[a]nthropogenic warming and sea level rise would continue for centuries due to the time scales associated with climate processes and feedbacks, even if greenhouse gas concentrations were to be stabilised' (Metz et al., 2007, pp. 12, 16). These sobering predictions underscore the immediacy of the clear and present danger of climate change. Since it is temperature increases that drive much of climate change it is appropriate to consider examples of the surface-warming scenarios from the IPCC report.

Some likely scenarios

A first group of scenarios (A1) is premised upon increased international integration. It envisions a future world in which regional associations become more important along with cultural and social interactions (Intergovernmental Panel on Climate Change, 2007b). International efforts to enhance the abilities of states to meet the needs of their citizens through effective and legitimate institutions substantially reduce regional differences including discrepancies in their per capita incomes. This scenario suggests that if the energy systems depended upon by states remain fossil fuel intensive, then sea levels could be expected to rise by 0.26–0.59 metres by the end of this century (Intergovernmental Panel on Climate Change, 2007b). Global temperature increases would be in the range of 2.4–4.6°C with the most likely being 4°C (Intergovernmental Panel on Climate Change, 2007b). In the unlikely event that non-fossil-fuel energy sources were globally adopted, sea level increases would still be in the range of 0.2–0.45 metres by the end of the century, with temperature increases of between 1.4 and 3.8°C and a best estimate of 2.4°C (Intergovernmental Panel on Climate Change, 2007b). In the worst of these scenarios, an increased global temperature of 4°C would lead to untold species extinction. Notable among them would be the loss of large whales whose breeding would be disrupted by climate change.

Most likely would be a situation in which states adopted a balanced range of non-fossil and fossil fuel energy sources. In this case, sea level increases by the end of the century would be in the range of 0.21–0.48 metres with temperature increases ranging between 1.7 and 4.4°C, with a best estimate of 2.8°C (Intergovernmental Panel on Climate Change,

2007b). Such a world would be significantly different from that of the 20th century. Rainfall patterns would have changed dramatically, resulting in wet tropics becoming drier with reductions in the amount of habitable land. Severe storms would have become common in many parts of the world along with more severe and widespread droughts. Wildfires and extreme bushfires would be more common, leading to further environmental pressures including increased airborne pollution and diminished forest resources. Severe flooding would occur across very large areas displacing millions of people and wreaking havoc on agricultural, industrial and social infrastructure.

The location of areas of intensive food production would also change as the effects of altered rainfall patterns and temperatures began to be felt. By the end of the century, existing farmland in many parts of the world would have become less productive while other regions might have increased their agricultural capacity. Animal migration patterns would have also changed with significant effects for food supplies based upon their harvest. As weather patterns forced animals to follow their habitats into new areas, further consequences would be experienced. Wildlife associated with the tropics, such as snakes, spiders and insects, would move into formerly temperate zones and many diseases and illnesses would find new patterns of geographic distribution.

In a second group of scenarios (A2) outlined by the IPCC, differences between states remain significant (Intergovernmental Panel on Climate Change, 2007b). Twentieth-century themes of self-reliance and the preservation of local identity would continue to shape state policies and global population growth would continue at relatively high levels. Economic development would be regionally oriented while growth in per capita incomes among states would be expected to differ. The climatic changes within this scenario are premised upon expectations that new technologies would be adopted by some states, but not all. By the end of the century, sea levels would be expected to rise by 0.23–0.51 metres with temperature increases in the range of 2.0–5.4°C degrees Celsius, with a best estimate of 3.4°C (Intergovernmental Panel on Climate Change, 2007b).

In the third (B1) scenario it was anticipated that rapid changes in economic structures would transform economies based upon industrial production into less polluting information and service economies (Intergovernmental Panel on Climate Change, 2007b). The development of clean and resource-efficient technologies would lead to reductions in intensive energy consumption. This scenario emphasises the roles of global solutions to economic, social and environmental sustainability, including increased equity among states. Yet even so, by the end of the century, sea levels would be anticipated to rise by 0.18–0.38 metres along with increasing temperatures in the magnitude of 1.1–2.9°C, with a best estimate of 1.8°C (Intergovernmental Panel on Climate Change, 2007b).

A further scenario (B2) emphasises the importance of local solutions in enabling future sustainability (Intergovernmental Panel on Climate Change, 2007b). As a consequence, the global population would continue to increase, albeit at reduced levels, and there would be a modest expansion in economic development. Diverse technological change would occur slowly among states although global interest would focus upon environmental protection and social equity. However, local and regional solutions would dominate at the expense of global responses to climate change. As a consequence, sea levels would be expected to rise by 0.18–0.3 metres by the end of the century and temperature increases would be in the range of 1.1–2.9°C, with a best estimate of 1.8°C (Intergovernmental Panel on Climate Change, 2007b). In all of these modelled scenarios, future rapid changes in ice flow, which would have significantly greater effects upon sea-level increases, were excluded.

Politics, geography and cooperation

The political dimensions of global climate change issues have exposed a mythical gap between the national and international activities of states. As a consequence of the established forms of modern state sovereignty, the limited scope and enforcement capacities of international law and the widespread acceptance of the principle of non-intervention in international relations, it has long been accepted that what a state does within its own borders is of little consequence to other states (United Nations General Assembly, 1970). However, this belief is now patently untrue and is rapidly becoming politically unsustainable. The effects of climate change do not respect the political cartography of the world and are experienced across state borders. Those activities which contribute to, or give rise to, such effects are rightfully of interest to the entire international political community.

Transboundary pollution and acid rainfall have already exposed many states to environmental damage from the industrial activities of neighbours. Regional responses have been formulated to mediate such effects by encouraging cooperative regulatory frameworks to minimise pollution and to compel states to alter their practices and/or provide compensation. By contrast, climate change and resource redistribution issues of the 21st century are global issues. They will impact upon all states in all aspects of their domestic production and international political-economic engagement. Addressing them will require much larger cooperative associations than bilateral or regional responses as have emerged in the past. It is likely that success will only arise from the development of new principles and modes of behaviour among states and other key international actors that provide universal guidelines of how human communities associate with the natural environment.

Re-evaluating international practices and accepted modes of behaviour in order to address the environmental challenges posed by climate change and redistributions in water resources also requires a reconsideration of the nature and value of sovereign territorial states. This becomes necessary because states are the central units of the international community. It has been suggested that:

> only a thoroughgoing ecocentric Green political theory is capable of providing the kind of comprehensive framework we need to usher in a lasting resolution to the ecological crisis... an ecocentric polity would be one in which there is a democratic state legislature (which is part of a multileveled decision-making structure that makes it less powerful than the existing nation state and more responsive to the political determinations of local, regional, and international democratic decision-making bodies); a greater dispersal of political and economic power both within and between communities; a far more extensive range of macro-controls on market activity; and the flowering of an ecocentric emancipatory culture.
> (Eckersley, 1992, pp. 179, 185)

Reforms of this magnitude are, however, unlikely to eventuate in a timely manner, and solutions will have to emerge from within, and accommodate, the existing international political community and its central actors. A fresh beginning clear of the impediments inherited from the past, such as a structural system of sovereign territorial states, is neither practical nor imaginable (Murphy, 1996). Nonetheless, it has been suggested that revolutions in the nature of state sovereignty are not only possible but a matter of historical record (Philpott, 2001). This gives rise to the possibility for new forms of political association, and attendant solutions to new issues, to be developed in an orderly manner.

There may be benefits to order arising from new patterns of development as these call for new forms of investment and technology transfer. A global shift towards renewable energy sources potentially improves states' energy security, although barriers such as the availability of appropriate technologies and finance, intellectual property rights, poverty and the complex uncertainties of economies in transition also need to be addressed (Metz et al., 2007, p. 20). Renewed demands for options such as nuclear energy are also within this mix of safety and security issues.

A political problem

States are embodied in governments and in many respects the terms have become synonymous. The central purpose of the state continues to lie in creating political order but also extends beyond territorial boundaries to include international political order. States cannot meet their domestic political obligations unless they also utilise foreign policy, diplomacy and other aspects of international relations. Much of what contemporary states do to afford protection to their citizens, and to guarantee their continuation, relies upon the relations they create with other states and the organisational structures they develop as instruments of international order. The political challenges of climate change, including the end of democratising processes and universal dreams of economic progress, require states and those who study their behaviour to reconsider the importance of sovereign territoriality, links between nations and states and governing principles.

Military industrial complexes and industrial development have contributed to environmental climate change issues in the process of securing the states system and state sovereignty (The Commission on Global Governance, 1995, pp. 29–30). Many of the activities that have secured and contributed to the maturation of the states' system of the 20th century, and the identities of modern sovereign states, have also created the problems of climate change, water resource security and political order that states confront in the 21st century. As observed in the United Nations Development Programme (2008, p. 27), confronting these threats:

> will create challenges at many levels. Perhaps most fundamentally of all, [climate change] challenges the way that we think about progress. There could be no clearer demonstration than climate that economic wealth creation is not the same thing as human progress.

The continuation of climate change impacts and their increasing challenges to human societies potentially jeopardise the further maturation of the current structures and mechanisms for order that characterised the contemporary international political community.

The implications and effects of climate change are likely to become sources and sites of new international tensions given the challenges being faced and their impacts upon all states. Unlike many other major sources of political realignment and international structural change, these challenges affect all states rather than particular regions. As Jackson and Sorensen (2007, p. 257) observe, these environmental issues potentially increase the risks of international conflict as is already evident in different regions of the world, such as Africa and the Middle East, where water has become an increasingly politicised resource.

New ideas and expectations

Those who study International Relations, in both theory and practice, have long had to concern themselves with global issues that do not conform to political boundaries. Transboundary issues such as shared waterways have concerned and impacted upon states since the earliest formulations of international law and have given rise to beliefs that states are interdependent. Solutions limited to bilateral, multilateral or regional scope are inadequate to address changes that are occurring at a global level. Unsurprisingly, such a tremendous upheaval in the international political community and its core components, the sovereign states, have challenged international relations analysts to recognise more fully the importance of complex interdependence (Keohane and Nye, 2001).

Ecological interdependence among states represents an important development in their relationships and requires new scholarly efforts to explain them, extending beyond collective security doctrines and analysis of multilateral agreements. The international political community of the 21st century is characterised by interdependent but stubbornly sovereign states. Order will be better maintained if new practices and structures 'are compatible with and... build on culturally familiar mechanisms of social control' that extend shared cooperation and support new practices of political responsibility (Biermann and Dingwerth, 2004; Young, 2002, p. 161). For instance, reinforcing the roles of international institutions might improve certainty in relations between states through increased relational stability.

The political issues arising from climate change are complex and states necessarily experience persistent uncertainties concerning their roles and priorities in cooperative solutions. Although the issues at stake in international climate change generally preclude comprehensive policy-making, this does not indicate that international policies are necessarily ad hoc. Environmental advocates, such as environmental NGOs and sometimes also key coalitions of states, can play crucial roles in establishing the parameters of negotiations, and the knowledge upon which negotiation and bargaining is premised. International environmental policy-making between a diverse array of actors is subject to their changing interests and abilities to influence agreements.

The complex nature of contemporary international relations is exacerbated by industrialisation, world trade and technologically advanced military power. At the end of the 20th century, economic decline within powerful industrialised states and rapid changes in the global balance of power – most notably the end of the Cold War and the collapse of the Soviet Bloc – contributed to an 'opening up' of the international political system. In the 21st century, increasing interdependency among states, both ecologically and economically, juxtaposes intricately complex problems of management and governance. These will achieve new intensity, require new levels of common political will and methods of reconciling the competing demands of human societies. It is now almost self-evident that one 'of the hardest lessons taught by climate change is that the economic model which drives growth, and the profligate consumption ... that goes with it, is ecologically unsustainable' (United Nations Development Programme, 2008, p. 27).

Conclusion

In the first decades of the 21st century, the developing states that 'missed the boat' of economic development in the 1960s and 1970s might prove pivotal in the world's capacity to respond effectively to pressing international environmental issues. As global climate change occurs, new power dynamics and efforts to maintain order are likely to produce alliances and cooperative endeavours that are intended to formulate effective strategies to deal with these issues. In the 1960s and 1970s, 'missing the development boat' resulted in enduring hardship for states that failed to achieve new forms of industrial and economic growth. Nonetheless, their experiences of economic hardship and persistent problems of poverty did not produce significant negative flow-on impacts for major industrial powers (aside from managing the debt problem). In the 21st century, if states 'miss the environment boat' in responding to climate change, energy and water resource issues, they will become unsustainable burdens for other states.

Many of those who study international environmental politics do not pay sufficient attention to the nature and capacities of key political actors, the structures of the international political community and the accepted practices by which states respond to international political challenges. The nature of states, their key characteristics and forms of existence are rarely addressed explicitly by those who examine the nature of environmental problems. This creates a gap in the literature and in our appreciation of political and diplomatic practices between states. It is taken for granted that all states share common assumptions and practices concerning the exercise and pursuit of sovereign authority while the nature of their national interests is observed to be diverse. If we are to understand the political importance of climate change and the redistribution of water resources, it is imperative that we attend to the features of states and the normative importance of their shared characteristics as sovereign entities.

It is often assumed that states have readily identified and targeted discrete objectives in all or most areas of domestic policy development. Such assumptions proceed from a general recognition of the structural components of states, their capacities to govern and the nature of legitimacy. Yet policy-making within states is rarely a simple or straightforward matter in areas such as industry, employment, energy and resources because they impact upon numerous economic and social goals and interests. In these areas, states take decisions that reflect their responses to tensions between short- and long-term interests, balancing choices concerning the distribution of benefits as well as developing strategies for implementation and policy review even though they identify policy objectives. The impending environmental crisis, presented by global climate change, is not restricted to particular states and does not recognise development status. The global ecosystems are shared by all, and, to that extent at least, the consequences of environmental damage of any kind constitute a global problem.

2
Limits to Global Consensus

Introduction

Global climate change is likely to make it difficult for liberal democratic political systems to govern effectively and to continue to set the rules of behaviour and leadership within the international political community. It is also likely to end the dreams of progress that liberal democracies have helped to produce. This does not necessarily mean that capitalism will fail, or that totalitarian regimes will flourish. It does, however, mean that contemporary liberal democracies will need to achieve effective new structures and identities as they adapt to an altered natural environment.

As global climate change unfolds, government leaders, citizens and other political actors will deal with new and confounding political realities (United Nations Development Programme, 2008; Eckersley, 2004). These changes will produce new rules for international political activities, including altered relationships between governments and their citizens. Governments and other international decision-makers will experience new conflicts concerning their diverse interests and will require new political structures to implement and manage their policies. Enabling human societies to survive these disruptions to familiar forms of order, security and stability and their visions of well-being will be important components of global climate change responses.

It seems reasonable to assume that governments and others accustomed to wielding power will respond to these uncertainties in a similar manner to their previous patterns of behaviour. We can, therefore, expect that governments will seek to create new economic policies and structures in the unfamiliar circumstances arising from less predictable weather patterns, changed rainfalls and the particular changes of seasonal temperatures and rising sea levels that affect them. Adapting to global climate change will require new forms and distributions of political power among human societies, and these will necessitate new forms of political organisation and produce new sources of political influence both domestically and internationally (United Nations Development Programme, 2008; Gelbspan, 2001).

Part of the reason for anticipating these changed relations between governments and people derives from the dramatic impacts of global climate change upon economic prosperity and the heavy reliance of many human societies upon the intensive consumption of natural resources. Extreme weather events, such as an increased incidence of severe storms, fires, heatwaves, floods, saltwater contamination of rivers and rising sea levels, will also alter citizens' expectations of their governments. These changed citizen–government relations are likely to be of such magnitude that they alter the distributions of rights and responsibilities within states and across the world (Biermann and Bauer, 2005; Young, 2002). Government reliance upon electoral popularity and sensitivity to complex networks of accountability may also change as climate change alters expectations of protection, security and recovery.

These realities will alter the ways that governments; security organisations; and humanitarian, emergency-relief and crisis-recovery teams support and protect human communities from devastation. Aside from the impacts of intense storms and floods, economic actors, such as banks, and insurance companies need to rapidly adjust their practices in response to altered distributions of habitable land, global food, industrial production and risk. Producers of electricity, plastic products and motor vehicles will become vulnerable to potential resource shortages, increasing demands for compliance with international agreements and specific domestic government targets. They will also remain sensitive to the economic pressures exerted by the end-users of their products.

Relying upon states

Climate change impacts will be unevenly distributed among states and some are likely to experience peculiar mixes of altered rainfall and temperature patterns (Parry et al., 2007). Altered climatic patterns will create unequal impacts upon human societies, all of which currently rely upon political structures and sources of sovereign authority for social order. Sovereign states will only remain the primary form of social organisation if the international community finds sufficient common ground in global concern for the effects of climate change. Identifying common interests in effective global responses will necessitate universal acceptance of states' continuing rights to supreme authority.

The history of liberal democracies coincides with expectations of economic prosperity and an array of rights among citizens. Throughout the 20th century, periods of economic decline produced dramatic swings away from claims to individual rights in favour of securing economic stability. For instance, the Great Depression of the 1930s contributed to the rise of fascism and other right-wing political forces within Western states. In similar vein, centralised forms of authoritarian government have been created by developing states seeking to secure prosperity in the face of continuing and apparently intractable economic problems. In both instances, however, these forms of government have shared in the 20th century dreams of securing economic prosperity and development.

While liberal democracies are often considered global leaders and major contributors to international peace, it is likely that these roles have arisen as much from the benefits of economic prosperity as from the benefits of representative forms of government and broadly based political participation supported by legislated rights. Indeed, over recent decades, it has been difficult to identify separate and distinctive political and economic visions, roles and status among Western liberal democratic states as their levels of internal pluralism and complex economic, political and social interdependencies have increased. For instance, it has been difficult to identify social imperatives as distinct from economic goals or issues of political representation from economic prosperity and beliefs in the benefits of a private-property-based capitalist economy.

The ideas, status and behavioural practices of states are now being challenged by new environmental imperatives. Global climate change impacts upon the distribution of human populations and their abilities to achieve viable forms of political association. Consequently, the institutions they maintain to support international political order will need to identify new sources of durable political authority. At the same time, states and their institutions will need to resolve new individual and collective uncertainties concerning the environmental impacts of economic production techniques and contexts regarding the security of resources. The magnitude of global climate change impacts are such that replacement technologies and alternative energy resources will not prevent rising sea levels from inundating cities, or rivers from drying as rainfall patterns alter dramatically (Dow and Downing, 2007).

Changing political parameters

These issues present major challenges for the international political community because individual states cannot readily act to mitigate the impacts of global climate change without making themselves vulnerable to changing political conditions. The international political community is comprised of structures and institutions that preserve the sovereign status of states. Throughout recent world history, efforts to achieve prosperity and international political order and stability have been ascribed priority over other political issues. Preserving states' rights to independent decision-making and interests has created an international political context that is poorly equipped for developing and implementing effective responses to global climate change. The structures and institutions supported by states prevent them from taking or initiating unilateral action and/or from seeking to drive rapid multilateral actions. Responding to global climate change requires states to negotiate their roles as central political entities and to adapt their structures and spheres of authority.

In the 21st century, the political challenges of global climate change are both too deep and too urgent for the international political community to continue to emphasise institutional and organisational consensus over changed economic and political practices. The key ideas of the 20th century promised greater economic affluence and broader distributions of political independence. These promises were believed to present achievable goals through state-based political authority in sovereign governments and were consolidated as states agreed upon their management of issues within their borders. In these and other ways, states preserved their status as legitimate decision-making bodies.

Presently the scientific consensus is that in the latter half of the 21st century, unchecked global climate change will cause sea levels to rise. As a consequence, among other effects, flooding will occur in the densely inhabited fertile Ganges River delta, the Thames River will become contaminated with saltwater and large parts of New York City will be inundated (Romero-Lankao, 2008; Nicholls et al., 2007b; Mirza et al., 2003). Over the same time frame, rising temperatures will make the affluent and previously temperate zones of Europe less comfortable locations for human habitation. States will then be challenged by a need to regulate population movements as people seek more conducive habitats. Maintaining productive capacities and stable populations will become difficult as people seek to take their forms of production and livelihood with them.

New forms of political organisation and human association will change the political structures and characteristics of states. Climate change 'refugees' from developed states will not wait until they have become poor, or hold fewer choices or diminished abilities to take capital and other potential sources of production with them. Developed states beset by climate change effects may simultaneously experience drains upon their human and resource capital that diminish their capacities to respond. These changing conditions will create levels of unpredictability in international political affairs, global production, investment and trade markets. As these processes unfold it will become increasingly difficult for states to find common political interests.

Presently, the international political community displays a level of complacency in relation to these issues. In the latter half of the 20th century, the international political community commonly practised consensus decision-making at major international forums where negotiations were intended to produce effective lasting agreements (Young, 2002). This followed the dominance of liberal democracies seeking to ensure that governments would follow through in implementing international agreements following the Second World War and the Cold War. Decision-making by international political consensus was perceived as necessary for limiting the impacts of lapses of political will as states' interests, long-term responses and strategies fluctuated. The histories of collective security arrangements, telecommunications agreements and law of the sea illustrate these dynamics.

In the 1980s and 1990s, international environmental agencies and intergovernmental authorities sought to implement agreements, such as the Montreal Protocol and moratorium on mining in Antarctica, and attempted to enhance compliance among parties. Consensus-based decisions offered favourable conditions for achieving these goals. In the 19th and 20th centuries, international norms and customary law emerged from customary practices which relied upon consensus for change and the implementation of agreed practices. These histories suggest that consensus has often been achieved through commonly accepted behavioural practices that have, in turn, established new conditions of trust and created positive decision-making settings (Koremenos, 2005; Mathews, 1991).

Structures, problems and uncertainties

The challenges of global climate change will require substantive adjustment to the practices, forms of association and institutional dimensions of the international political community. The abandonment of consensus style decision-making will be among the changes required, and this creates new political uncertainties because consensus decision-making has been valuable in producing effective lasting agreements (Young, 2002). Nonetheless, as global climate change creates new political uncertainties consensus appears unlikely to be appropriate for a number of reasons. First, a consensus approach is time consuming as successive rounds of negotiation slowly bring divergent parties together. The scientific community is increasingly certain that we do not have the time for such an approach. Second, international environmental agencies and intergovernmental authorities are likely to seek higher levels of compliance among parties by adopting penalty clauses. Third, as is evident from the 2009 Copenhagen Climate Change Summit, the magnitude and diversity of the challenges confronted by contemporary states, coupled with their long histories of economic and political competition, make it unlikely that they will readily achieve meaningful consensus in the near future.

The consensus approach, however, may not be easily abandoned by states. Major consensus-based international agreements provide tangible evidence of their possible attainment and many international actors believe that durable agreements rely upon these decision-making processes. In the past, consensus-based decision-making proved beneficial as various international agreements provided frameworks and structures: conventions concerning the rights of children, transboundary and marine pollution agreements provide salutary examples. Throughout the history of the modern world, changing practices and behaviour among states formed the basis of international law and consolidated the central political status of states.

Addressing global climate change will challenge existing forms of human association and the decision-making processes that currently provide the basis of orderly conduct. Timely and grand-scale responses are now widely believed to be necessary for ensuring sustainable societies and these will alter many dimensions of human social organisation and give rise to debates about the costs and benefits of change. Arguments concerning the nature and extent of sovereign responsibilities increasingly have environmental components and suggest that those who neglect global climate change responses, risk endangering their legitimacy and ongoing influence within the international political community (Mathews, 1991). Changes in the dynamics of the international political community suggest the 'shadow of the future' has shortened to reveal the natural environment as an ongoing site of contested authority between states (Bearce et al., 2009; Donnelly, 2000; Paterson, 2000). Under the influence of this shadow, states, intergovernmental organisations and private actors are grappling with new uncertainties regarding their authority and decision-making capacities. For many actors, these uncertainties extend to their perceptions of interests and priorities as they seek to accommodate new knowledge and reconcile changing expectations as predicted levels of threat alter their goals.

The failed 2009 Copenhagen Climate Change Summit revealed two important political dynamics as additional factors in international climate policy initiatives. First, there are mixed patterns of awareness and denial in the international political community about the effects of global climate change. Second, there are mixed patterns of awareness and denial among actors in the international political community about the magnitude of political turmoil that contemporary human societies confront as they consider potential responses to global climate change. How to respond and adapt to global climate change has become the key to future debates (United Nations Development Programme, 2008). Finding ways of coping and adjusting practices, lifestyles and forms of association now pose challenges for states that are accustomed to institutional stability as sources of effective political and economic organisation.

Mixed results of sovereignty

When states participate in international negotiations concerning climate change and related environmental problems, they confront very real difficulties in reconciling the competing demands and imperatives to which they are subject. Their need to protect citizens, ensure economic prosperity, preserve territorial security and reserve their status as recognised legitimate authorities intersect in ways that make it very difficult for them to discern their specific environmental interests. These same imperatives also limit their abilities to respond to grand-scale external problems, such as rising global temperatures and changes in global weather patterns. In this context, sovereignty provides states with an internal compass. It provides them with common means of claiming and retaining their status as primary political authorities, and it is to this touchstone that states repeatedly return.

State sovereignty continues to provide an important source of stability, despite the fact that transnational issues repeatedly defy individual state resolution (Holsti, 1995). It is easy to be critical of international inaction on climate change issues, such as the slow pace at which the

Kyoto Protocol was ratified, and the persistent refusal of some states to join it. However, the diverse nature of states, and their inherent levels of competition in maximising their access to resources and security, makes it remarkable that consensus style actions ever occur. Global climate change highlights political fragmentation as a symptom of the limited abilities of the international community to address problems that 'rarely correspond with state boundaries' (Holsti, 1995).

Alongside the growth of modern political economic arrangements and expectations of progress, sovereignty has been premised upon expectations of the legal status, authority and decision-making capacities of states. It has sat alongside visions of international order and stable relations between states, a distribution of power among political actors and assurances of progress. In the 21st century, as global climate change requires states to develop new levels of flexible responsiveness to establish effective adaptation strategies and altered forms of human association, sovereignty has become part of the problem for the international political community (United Nations Development Programme, 2008). By individuating interests among states, sovereignty inevitably favours domestically located interests over a collective good – except in rare circumstances of consensus-based collective will (Gleeson and Low, 2001). As states respond to these pressures and adjust to their evolving status and capacities, opportunities for consensus will diminish – at least temporarily.

The dilemmas arising from international regulatory efforts present states with new decisions concerning their status as political authorities and sources of collective identity for their citizens. In the 20th century, states' interests emerged from perceptions of their own interests as largely existing outside the realm of global order (Edmondson and Levy, 2008). Knowledge-based actors of various kinds, such as scientists, intergovernmental organisations and non-governmental organisations sought to influence states' perceptions of the impacts of greenhouse gases in the latter stages of the 20th century. They also sought to understand the importance of gaps and tensions between international and domestic interests because they perceived political structures and diverse interests to be important components of the economic, social and political undertakings of states (Paterson, 2000; Young, 1989). Thus, we can now appreciate the roles of knowledge and policy networks as 'uncertainty reducers' that play multi-layered roles in coordinating 'policy arrangements' (Haas, 1995, p. 4).

Throughout the 20th century, it became possible to identify conditions under which consensus could be promoted (Sebenius, 1992;

Haas, 1990). For instance, trust among parties, low risks of free-rider behaviour and readiness to overcome small logjams in negotiations over specific issues all create favourable conditions for negotiated consensus (Young, 1994). Similar dynamics occur when the costs of participating in establishing new international agreements are low and/or the costs of accepting the behavioural constraints specified in an agreement are minimal. By contrast, high-intensity interests and decisions that impact directly upon domestic economic and political interests decrease the likelihood of international consensus (Kütting, 2000).

Optimism and consensus

In the late 20th century, heightened optimism concerning international consensus-based decision-making followed the successful establishment of the Intergovernmental Panel on Climate Change (IPCC) and the ready implementation of the Montreal Protocol on Substances that Deplete the Ozone Layer targets (Oberthür and Ott, 1999). The stalling patterns surrounding the Kyoto Protocol and its protracted difficulties in achieving wide acceptance among major emitters of greenhouse gases reflect the difficulties faced by the contemporary international political community in achieving even small steps towards effective climate change responses. States are predisposed to favour 'next time' dynamics, delaying decisions in the hope that further information or changed scenarios will make actions unnecessary (Kütting, 2000; Scharpf, 1989). When the IPCC was formed, and the Montreal Protocol agreed, they benefitted from the early momentum associated with an initial wave of urgency that positively contributed to 'we must do something' international political will.

More recent obstacles to consensus have revealed global climate change issues as resistant to programmatic decision-making. The need to accommodate diverse interests and new knowledge increases resistance as changing political conditions of global climate change reduces the likelihood of consensus and reduces the quality of agreements. When consensus is achieved, it will be likely to coincide with expectations of low levels of compliance, weak enforcement mechanisms and/or politically straightforward issues.

Political will is a vital component of international collective agreements for responding to climate change impacts but achieving common goals and developing cooperative strategies for adaptation and mitigation is not guaranteed by it (Young, 2002; Vogler and Imber, 1996). New political challenges to the nature of sovereign states and the forms of economic and social association they create are likely to transcend matters of commitment, resource allocation and states' abilities to enact new legislation that supports international agreements and their compliance requirements (Postiglione, 2001). Where states have made good progress in responding to potential impacts of climate change, these positive actions tend to have occurred when domestic political interests could be achieved through new policies or sector located initiatives to benefit an economic sector. An example of such a success can be identified in the growth of a sugar production/ethanol fuel industry alongside a hybrid motor vehicle industry in Brazil (Hoffman and Hoffman, 2008).

Changing interests and status

Climate change consequences cannot be isolated to the interests of individual states, groups or coalitions of states sharing interests. Initiatives that homogenise interests among diverse states are also unlikely to deliver meaningful outcomes (Gupta, 2005; Haas et al., 1993). Responding to climate change consequences successfully will depend more upon the functional aspects of sovereign statehood and less upon their shared legal status as central political actors (Biermann, 2005; Postiglione, 2001). This will challenge the international political community because it presents a radical revision of its structural dimensions since legal status through recognised equality has, to date, held primary consideration. Climate change consequences therefore highlight the extent to which sovereignty constitutes a 'magic mirror' through which states' identities are not merely reflected but are formed, consolidated and revised.

In the 1980s and 1990s as the first major international revelations of climate change were debated, states confronted simpler choices. They could behave as 'world leaders' by participating in international agreements, creating new authorities, agencies and organisations. They could decide not to participate in such negotiations, but reserve the right to later join decision-making forums. In the 21st century, as climate change debates have grown more complex (in tandem with the identification of more complex consequences), it has become more difficult for states to make clear assessments of the relative merits, costs, burdens and benefits of participating in specific conferences, conventions and treaties (DiMento and Doughman, 2007b; Biermann and Bauer, 2005).

As states grapple with changing expectations of shared responsibilities and mutual interdependencies in seeking to respond to global climate change impacts they must also recognise that changes to their physical environments inevitably redistribute sources of authority. These factors change the relationship between the various social and economic sectors over which states exercise authority. As states become more reliant upon specific knowledge and problems identification, they will produce new forms of commitment to the specific interests they represent and seek to protect (Thompson, 2001; Haas et al., 1993).

The range of environmental problems that became politically salient in the 1980s simultaneously challenged notions of national sovereignty and reinforced interdependent relations between states. These environmental problems tended to stress the interdependence of states, highlighting biological and geophysical factors in supporting human (and other) life, and presenting new demands upon international political systems and structures. As Mathews (1991, p. 31) observes:

> *The nature of sovereignty is changing.* The global environmental trends, loss of species, ozone depletion, deforestation on a scale that affects world climate, and the greenhouse effect itself – all pose potentially serious losses to national economies, are immune to solution by one or a few countries, and render geographic borders irrelevant. By definition, then, they pose a major challenge to national sovereignty.

Engaging with some of the international political dimensions of global climate change and the manner in which it raises new questions regarding the nature of rights and responsibilities and their distribution will be an essential component of 21st-century climate change responses.

Pessimism and responsibilities

As global climate change challenges increasingly dislocate people from their former homelands and alter the relationship between people and their states, and change patterns of economic production, states' abilities to claim, delineate and reiterate their specific interests will become increasingly contested (United Nations Development Programme, 2008; Paterson, 2000). The dream of global consensus cannot be preserved under these conditions. This does not mean that a descent into international anarchy is likely or inevitable, but rather that the power of states remains limited, notwithstanding their abilities to retain primacy as international political actors. States' abilities to claim uncontested rights to exert political authority will diminish as specific climate change impacts become more easily identified in local contexts, thereby challenging their central political institutions. It is likely that states will endeavour to reassert their superiority over other political entities, albeit with expanded concessions to their collective responsibilities (Archibugi, 2001; Paterson, 2000).

In the modern international political system, the abilities of states to govern, produce and trade depends upon the availability and management of an extensive list of resources. Maintaining political order, sovereignty, security and prosperity are intimately aligned with resource use. Global climate change renders international consensus too slow, and insufficiently reliable, to secure visions of global democracy in the 21st century. As climate changes impact unevenly, a state that cannot demonstrate its acceptance of responsibility for preserving commons resources might expect to find its legitimacy questioned.

Reconciling tensions

It is difficult to resolve debates between and among states concerning who holds responsibilities for progressing climate change responses and who should bear the costs of their implementation (Matthew, 2007; Roberts and Parks, 2007; Achterberg, 2001). There are two parts to these debates. First, there is debate centred upon the relative distribution of burdens and responsibilities among states and economic actors based upon their past and present contributions to climate change. In these, there exists a crude division between developing and developed states. The developing states see themselves as further disadvantaged and marginalised by new international efforts to regulate the technologies and industrial practices they desire (Roberts and Parks, 2007). They believe the greater burden should be shouldered by developed states. Not surprisingly, the developed states who have been major contributors of greenhouse gases see the situation differently. They argue that climate change is a global or collective problem that can only be effectively addressed with all states working and making sacrifices together (United Nations Development Programme, 2008; Roberts and Parks, 2007).

Second, focusing upon global responsibilities, rather than apportioning blame for the emergence of climate change, will enable the international political community to transcend the limits of interests-based claims to international participation. It will enable states to reconfigure their authoritative status in order to achieve a systematic evolutionary solution to a structural problem that poses serious threats to the political systems, societies and people of all states. As international climate

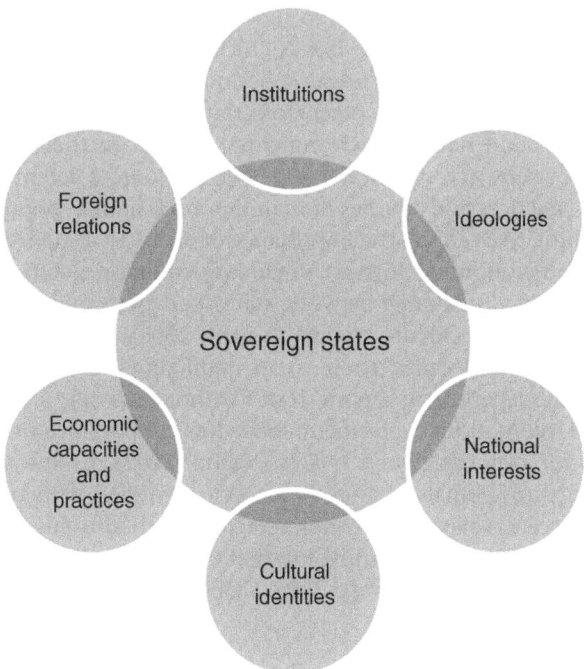

Figure 2.1 Sites of diversity among states

change scientists and policy experts increasingly point out, the more we learn of the potential challenges arising from global climate change, the more we realise we have yet to learn (Parry et al., 2007). The particular complexities of climate change and its anticipated impacts upon the distribution of water resources, viable agricultural land and land mass suited to human habitation mean that international political agreement concerning how to respond to these challenges will be an important component of any mitigation efforts. However, these very same features also increase the difficulty of attaining international agreements (Figure 2.1).

States rely simultaneously upon their functional capacities and status as sovereign authorities (Edmondson and Levy, 2008). Their physical environments impact upon their abilities to exert authority, secure their borders and pursue economic prosperity and indirectly determine their abilities to provide rights and protections. The unfolding consequences of global climate change provide salutary reminders to all states that their functional capacities do not automatically arise from their sovereign authority (although this is crucial to their abilities to maintain domestic social and political cohesion and to engage in orderly international dealings). Even prosperous states depend upon their physical resources and natural environments to maintain their functional capacities because access to water and energy resources determine their industrial practices and abilities to maintain their human populations.

Continuing to overlook the importance of diversity in rights, responsibilities and notions of interest (including common good) can only result in further inequities between states and perpetuate ineffective responses to global climate change (Najam, 2005a, p. 245; Bonanate, 1995). For several decades, these issues highlighted levels of inherent physical interdependence among states without making it any easier to achieve international agreement about their importance or effective strategies for managing their effects (Young, 2002). As the potential consequences of climate change have been identified with increasing clarity in recent years, key economic actors have also engaged more fully in public debates concerning international and domestic economic policies (Arnold, 2011; Flannery, 2005; Gelbspan, 2001).

Enduring weaknesses

A key weakness in sovereignty as an organising principle for relations between states, and their identities as supreme political actors in world affairs, lies in the fact that it does not create functional or real equality between states. Instead, it provides a façade through which unequal forms of 'equality' are maintained and through which the independence promised by sovereignty has been something of a holy grail for modern states, especially those shaped by liberal institutional ideals (Edmondson and Levy, 2008). This has created a so-called 'equality among states' which has been promoted, supported and reinforced by forms of collective decision-making (Hoffman, 1997). The localised impacts of climate change and their challenges to states' rights of autonomy will highlight the extent to which this theoretical legal equality contributes little to the practical, functional dimensions of governmental authority as not all states exert supreme authority over the same range of issues (Matthew, 2007).

Conclusion

States have been accustomed to individually determining areas of political activity over which they require supreme authority. They have also held rights to select areas in which they might share or delegate authority to other institutions when such delegations have supported their specific interests. However, global climate change impacts are dramatically altering states' abilities to claim and exert such rights while simultaneously extending their fields of shared jurisdictional responsibilities (Postiglione, 2001; Bonanate, 1995). The decisions they are required to make, and the responsibilities they must assume in order to avoid the worst case climate change scenarios, will negatively impact upon daily activities in all societies.

In the 1980s and 1990s, umbrella-style agreements and comprehensive regimes were relatively easily negotiated (Hempel, 1996). States could comfortably participate in international discussions and secure their enduring rights to international political participation and decision-making by involving themselves in multilateral collective action agreements. Such agreements only rarely required specific action. More commonly, they enabled the establishment of new international agencies or authorities to act on behalf of parties, or they entailed a step in an iterative agreement among limited parties (such as in regional environmental agreements) (Hasenclever et al., 1997). In the 21st century, these forms of international agreements have been revealed as poorly equipped to deliver effective responses to climate change challenges for the changed international political community.

It is precisely because of their status as primary political actors that states must now assume new responsibilities in public education and demonstrate leadership in establishing new linkages between the economic and political spheres of their diverse human societies. In short, it is because states hold legislative power and authoritative capabilities that other political, economic and social actors see them as responsible for demonstrating political leadership in responding to global climate change (United Nations Development Programme, 2008; DiMento and Doughman, 2007b). Effective climate change mitigation and adaptation strategies will therefore rely upon international law with supporting domestic legislation enacted by member states (Oberthür and Ott, 1999; Rittberger, 1995).

The unequal impacts of climate change consequences will further increase and complicate existing patterns of diversity as resources, such as rivers, agriculturally productive land and species cease to exist in some locations (Dow and Downing, 2006; Dauncey and Mazza, 2001). As some parts of the globe become wetter and some become drier patterns of human life, its forms of social organisation and relations between states and their citizens will also be altered (Barry and Eckersley, 2005). This creates a set of political problems for the international political community because alternative forms of human association and political organisation are not yet established and these are unlikely to be readily agreed by all political actors. Responding appropriately to climate change consequences involves examining the characteristics of 20th-century globalisation that generated new and more effective means of accommodating states' diverse rights, responsibilities and interests.

Climate change poses genuine challenges to current ideas about how human societies are politically organised and how these societies configure themselves as sovereign states as they pursue their interests within the international political community (United Nations Development Programme, 2008; Eckersley, 2004). As political and economic practices are altered by climate change consequences, it is likely that established patterns of international negotiation and the norms of international relations might also be altered. As the impacts of climate change unfold in the 21st century, states' practices and expectations of achieving their political interests through international diplomacy, trade and collective security arrangements will also be altered. These changes will be expressed in the ways states seek to represent and extend their interests, including in the ways they seek economic prosperity and attempt to secure their rights to political participation and representation within international organisations.

3
Governing Nature and Global Governance

Introduction

Global climate change challenges many of the structures, processes and practices of the international political community and threatens the key political visions, beliefs and practices that underpin the roles and capacities of core actors. As argued in Chapter 2, the political structures and institutions arising from consensus-based decision-making processes will not achieve effective, efficient or timely international climate change responses (Stern, 2009). While slow responses sometimes arise from a lack of political will, it is more often the case that policy-makers become caught in uncertainties concerning risk assessments and fluctuating prospects of effectively managing changed practices (Young, 2002). The political impacts of global climate change go beyond the consequences they pose for the forms, locations and distributions of human societies and their centres of production. They challenge core political values and ideals that seem fitting for the most significant set of issues yet to have faced human civilisations. Problematically, the prospect that existing core political values are challenged by global climate change is a dawning realisation that few political actors readily accept and acknowledge.

We can no longer afford to overlook or ignore the importance of political structures and institutions to achieve effective mitigation and adaptation strategies. The levels at which greenhouse gas emissions targets are set, and whether or not they are achieved, and/or are regarded as tradable commodities are important to how greenhouse emissions will be reduced and how global energy production continues (Stern, 2009; Hoffman and Hoffman, 2008). Identifying who will support or limit such initiatives presents additional political issues that also require urgent attention. Climate change policy outcomes will be affected by the status and capacities of those who enter into international agreements and the means by which they follow through in seeking to 'keep their promises' (O'Neill, 2009; Young, 2002).

Effective responses to mitigate the worst consequences of global climate change rely upon global governance mechanisms that commit states to more responsible courses of action. Achieving these will rely upon collective agreements and effective authorities to ensure that internationally agreed targets are met. Orderly and equitable mitigation strategies cannot simply be achieved through broadly based agreements concerning shared responsibilities and common goals of ensuring the longevity of human societies. Such goals necessitate changed international approaches to economic and social policies, including new structures for resolving contested interests (United Nations Development Programme, 2008; Najam, 2005b).

Incorporating ethical approaches into international decision-making and state-based authority structures will be an important component of effective global climate change responses. New political visions and international structures rely upon new ethical principles including new means of determining the legitimate status of states and other international structures (Eckersley, 2004). Lengthy re-negotiations between diverse actors will be required to enable joint agreements and the adoption of new practices. Like all major transformations in international politics, these changes will occur over a long period, emerging through the adoption of new customs and accepted behaviours (Edmondson, 2009; Nelson, 2009). This is part of the political context within which individual states and the international political community confront the challenges of identifying new ways of managing their rates of energy consumption, methods of production and environmental exploitation.

Creating the political conditions under which global governance might be achieved requires continuing persistence by those best equipped to demonstrate political leadership and patience among those whose lives are beginning to be directly impacted by climate change. The international political community is not without precedent in establishing collective means of managing environmental problems that cross territorial boundaries. However, many of the agreements currently in place impact upon only limited sectors of production within particular states, or require modest supporting domestic legislation. Additionally, many of these agreements were formed by relatively small groups of states seeking to address regional problems (O'Neill, 2009).

Current visions of societies and the beliefs they hold concerning their individual and collective means, rights and prospects of order and prosperity are challenged by global climate change (Kütting, 2000). The political consequences of global climate change are both deep and complex, rendering a great many aspects of contemporary economies, patterns of human settlement, production and overall lifestyles subject to disorder (United Nations Development Programme, 2008). These effects are not limited to matters of production, energy consumption and health-related consequences but go to the heart of widely held beliefs concerning the relationship between present and future societies, viable forms of human communities and the roles of natural resources in securing their longevity (Matthew, 2007; Thompson, 2001). Presently, various actors grapple with particular climate change–related challenges, and their efforts are admirable. However, many of these tend to miss the central point that addressing global climate change requires the contemporary world to relinquish particular political visions and simultaneously generate new visions to support viable futures. This is an intensely difficult project.

While global climate change poses significant challenges for political, economic and social authorities, global governance mechanisms can support their efforts to manage, mitigate and adapt. Redistributions in water and other resources, changed availability and access to viable agricultural land and the range and distributions of species will present new complexities in political decision-making and visions of community well-being. These new visions, processes and structures must take account of the nature of human interactions with their localised natural environments, which challenges an international political community premised upon progress, democracy, rights and prosperity (Edmondson, 2011; Eckersley, 2004). Shared global governance mechanisms and aspirations will support altered relationships between human societies and their geophysical environments. Only global governance can galvanise, coordinate and sustain universal responses to global climate change.

The problem with progress

Established political structures and visions of political and economic progress create obstacles for meaningful international climate change responses. They limit flexibility and creative problem-solving by reinforcing the centrality of prosperity-related goals and require political decision-making to prioritise economic interests. Global climate change exposes the myth in these views of progress and human control over the natural environment (Hughes, 2009; Low and Gleeson, 1998). In this way, the political ramifications of global climate change challenge the foundations of modern human societies.

At least since the Scientific Revolution, Western societies and the governments that have been granted authority over them have believed it both possible and desirable to exert human control and regulation over the natural environment (Barry and Eckersley, 2005; Paterson, 2000). This process did not begin with the Scientific Revolution: human settlements have long sought to transform landscapes for agricultural purposes and control aspects of the environment to improve human habitability, such as in interrupting or regulating the flow of rivers for irrigation (Hughes, 2009; Flannery, 2005). However, the discoveries and knowledge of the Scientific Revolution also encouraged perceptions of humans as readily capable of managing the environment. They created new levels of certainty for agricultural production, expanded mining operations and underpinned industrialised production processes. Over time, these practices allowed individual producers and their broader societies to maximise harvests and more efficiently utilise the natural resources to expand production and lifestyles.

Global climate change impacts are beginning to compel human societies to appreciate the extent to which their visions of social and economic progress have rested upon beliefs in human abilities to control or regulate the relationships between themselves and the natural environment (Barry and Eckersley, 2005; Thompson, 2001). The political choices faced by states in addressing global climate change are rather more complex than many advocates of alternative future visions suggest. They are not limited to finding alternative clean energy sources, or preventing the growth of new industrial centres in developing states. It is not a simple matter of finding new technologies to replace old ones, such as cleaner sources of high-capacity electricity production. The heart of the issue lies in finding new political visions and structures to support new economic systems and production practices that are less environmentally disruptive (Roberts and Parks, 2007; Kjéllen, 2006; Barry and Eckersley, 2005).

At no other time in human history has it been necessary for so many diverse political actors to agree upon such complex sets of issues. Responding to the political, economic and social challenges of global climate change requires states and those who seek to influence their regulatory and policy choices to reconsider how they provide for social order, how their societies are organised, what they produce and how they engage in continuing production. The conventions concerning

Law of the Sea, the manufacture of ozone-depleting substances and transboundary pollution agreements provide examples of earlier international agreements on environmental matters. In these instances, fewer states were engaged in decision-making and the issues at stake were more contained, at least in terms of their importance for domestic social and economic sectors.

The problem with democracy

Global climate change confounds the abilities of individual states and the international political community to maintain their beliefs in sovereignty as the central pillar of international political organisations. These new political demands make it unlikely that continuing to adopt democratic decision-making and related political processes will produce good outcomes in international decision- and policy-making forums. In attempting to maintain democratic decision-making, states are unlikely to achieve timely responses or produce agreements that incorporate appropriately stringent or ambitious targets (Edmondson, 2009; Gallagher, 2009). This would be less problematic if multilateral agreements were more likely to be equitable, durable or effective. Unfortunately, this is not the case.

Although liberal democracies remain a numerical minority, they have exerted strong levels of influence over the structures, institutions and established practices of the international political community. Notwithstanding the validity of debates concerning their roles in constructing an international system, best characterised as a Western hegemony, the international political community that now grapples with climate change is a product of the political and economic visions and practices of Western democracies (Roberts and Parks, 2007; Eckersley, 2004). While this is partly a consequence of the economic prosperity they have enjoyed (which has enabled their material domination over less powerful states), it is also a consequence of their compelling political visions over the last couple of centuries (Roberts and Parks, 2007; Barry and Eckersley, 2005; Haas et al., 1993).

Capitalist economies based on private property ownership have sat alongside beliefs in progress. Entrepreneurial economic activities have sat alongside individual freedoms. None of these arose as mere accidents of the Industrial Revolution but rather as specific political values encompassed within visions of particular forms of society. They included ideas concerning the means of progress and certain roles and relationships between people and governments. When states make decisions concerning their domestic policies, they do not separate their political and economic interests because citizens are simultaneously economic producers and consumers and participants in political activities. In democracies, those who obey the rule of law are also those who form and administer government, and then hold governments accountable for their actions. In taking and implementing decisions, governments rely upon financial resources drawn from the people and industries through direct and indirect tax contributions.

New technologies and dilemmas

Greater levels of global governance will be required to enable meaningful mitigation and adaption responses to climate change. It is now widely accepted that both urgent mitigation strategies and concurrent adaptation strategies will be required and that these must be broadly distributed worldwide (Stern, 2009; United Nations Development Programme, 2008). Precisely what these entail and who should produce and maintain them remains contested. We share Stern's view that mitigation and adaptation responses must be concurrently implemented, although we do not share his optimism regarding ready technological interventions that might minimise climate change–related disruptions to economic prosperity and a broader distribution of affluence throughout the world (Stern, 2009). This is partly because we do not share Stern's beliefs concerning the relative ease with which states will accept levels of political and economic interdependence that they have sought to limit for several centuries. Even if new technologically based solutions to global climate change could be readily identified, the time frames required for production and the scale of infrastructure adaptation or replacements limit their rapid uptake.

New technologies do not yet offer guarantees of increased efficiencies or other gains in the consumption of energy or water supplies to make their adoption compelling. Neither do they guarantee enduring benefits to climate change management. Those who anticipate that new technologies will provide salvation from the problems of climate change attribute a degree of rationality to technological solutions that has not been warranted to date. Many new technologies present new and potentially ongoing problems as their fuller impacts become apparent (Gallagher, 2009; DiMento and Doughman, 2007; Dow and Downing, 2007). For instance, low-energy light bulbs use less electricity but create new challenges for retrieving and recycling the mercury they contain.

In the contemporary international political community, technologies do not transform social preferences and they are not evenly or equitably distributed across the world. Markets do not rationally determine production or consumption and their influences upon the structures of societies are unclear (Ellwood, 2009). The diversity of approaches among states in addressing their energy requirements provides just one instance that demonstrates how technologies are transformed in their development and use by pre-existing socially derived preferences. This is likely to remain the case as alternative energy sources are developed in response to climate change and will mitigate against uniformly adopted solutions (see Chapter 7 for further discussion).

In so far as technologies will contribute to climate change mitigation and adaptation efforts, there must first be broadly based acknowledgement of the issues to be addressed and shared commitments to the nature of the solutions. Adopting new technologies would require rapid global availability, which would rely upon changed production and distribution networks. Ensuring that new technologies are available to those who need them but are unable to afford them would rely upon targeted subsidies that are internationally managed and equitably distributed (Roberts and Parks, 2007). Achieving material equity among states would require unprecedented levels of global organisation and sustained political commitment (Roberts and Parks, 2007; Gupta, 2005). These realities make the adoption of technological solutions subject to political processes alongside engineering and economic considerations.

The problem with economic order

At least since the imperial activities of Western European powers colonising Africa and Asia in the 19th century, almost the entirety of the globe has shared, to greater or lesser extents, beliefs concerning economic prosperity as a cornerstone of human happiness and the longevity of societies (Edmondson and Levy, 2008; Hoffman, 1997; Biersteker and Weber, 1996). This has been linked with a belief in the relationship between people, governments and time, such that shared views concerning progress as a social good have been cultivated, notwithstanding some accommodation in differences of opinion regarding the specific characteristics of such progress. At present, states hold different views of desirable and appropriate forms of political organisation. However, all states perceive themselves as holding authoritative rights to solve social problems within their territories according to their specific preferences (O'Neill, 2009; Edmondson and Levy, 2008).

Table 3.1 The primary decisions of states

Determinants of international activities	Determinants of domestic governmental activities
Who will be our friends?	Who will we include as citizens?
Will we hold alliances? With whom?	What rights will our citizens hold?
Will we trade, with whom and how much?	Who will make which government decisions and how?
What will we trade and how will we protect our markets?	What is the relationship between the social, economic and political sectors?
How will we protect our territory, people and institutions?	Who is responsible for the distribution of benefits and advantages and what do these entail?

By and large, modern governments and the people over whom they exert authority display ready acknowledgement of progress as an indication of good government. They hold a common belief in the importance of governments in creating and maintaining social order, regulating levels of conflict and enabling trade through labour, production and market formation and maintenance. They share a belief that good government can extend the life expectancies of people, create conditions in which material comforts can increase and produce an all-round sense that life today is better than in times past. These perceptions are evident in the ways that states regulate and legislate, including in the ways they relate to other international political and economic actors (Table 3.1).

The expectations and opportunities created by intergovernmental organisations and the international political community influence the policy directions pursued by many states. Throughout the 20th century, liberal democracies displayed international political leadership in adopting collective security arrangements, and structuring the international political community to support orderly conduct between states. In the latter decades of the 20th century, these principles were extended to include humanitarian intervention, which in some respects provides a template for recognising collective responsibilities for developing and implementing climate change mitigation and adaptation strategies (Edmondson and Levy, 2008). Even as climate change consequences directly threaten prospects of prosperity, the international political community is still coming to terms with this changed political context (United Nations Development Programme, 2008; DiMento and Doughman, 2007a; Roberts and Parks, 2007).

Historically, states and the international political community have maintained a conceptual and policy gap between their economic and security interests (O'Neill, 2009). They have considered their economies in terms of capacities for growth and prosperity and the relationship between states and citizens (whereby citizens produce, consume and participate in the allocations of resources across their societies). In their international relationships, states have been more directly concerned with preserving their territories and independent rights to exercise domestic legislative capacities (Edmondson and Levy, 2008). These conditions are changing as global climate change threatens economic and security interests with little regard for whether these are primarily domestic or international (Kütting, 2000). Interplay between trade, security, prosperity and international leadership interests among states complicates their assessments of risks and stifles their decision-making especially when electoral political sensibilities are at stake (Matthew, 2007).

Political values, which have long driven government regulation and policies, must now also drive international regulation and policy directions (Biermann and Bauer, 2005; Low and Gleeson, 2001). We believe this can be achieved because the long history of shared recognition of sovereignty enables the present international political community to apportion rights and responsibilities to states. Sovereignty provides states with privileged status as political authorities and, thereby, both permits and demands their participation in legislative activities, international trade and territorial defence. Historically, these legislative and regulatory activities have largely focused upon domestic interests, but interdependencies of global climate change are extending states' international interests.

In seeking to understand international climate change–related discussions, it is important to bear in mind that state-based practices and beliefs in autonomy and sovereign independence are not values neutral. Expectations about independent and autonomous governments carry a deeply ingrained acceptance of ongoing competition among states. Many expect that such competition is a permanent feature of relations between states creating 'us' versus 'them' dynamics that sustain state identities beneath the veneer of their collaborative cooperation (Doty, 1996). Global climate change, therefore, challenges the entire international community, creating new patterns of political and economic winners and losers.

Building political leadership

In the 20th century, liberal democracies demonstrated political leadership in establishing, maintaining and refining the (admittedly imperfect and sometimes oppressive) institutions and structures of the international political economy (Edmondson and Levy, 2008; Haas et al., 1993). In these, and myriad other ways, liberal democratic states influenced the behaviour of other states, including by establishing norms of accepted behaviour and supporting the development of international law. They also influenced the goals, functions and capacities of international governance mechanisms and set parameters for their legitimacy. In some respects, this is unsurprising since the liberal democracies tend to be industrialised states whose status as leading international political actors dates, one way or another, to the establishment of modern sovereign states.

In the contemporary international political context, the liberal democracies perceive themselves as guardians or trustees of order and security. They are engaged in developing shared recognition of collective responsibilities in extending human rights and protecting citizens from various threats. Their political institutions reflect their histories of liberal democratic ideals and progress-fixated economies that have emphasised rights as fundamental to good government and privileged economic growth over other social benefits (Barry and Eckersley, 2005; Bonanate, 1995). As guardians of dreams and promises of prosperity and progress, liberal democracies have led other moments of international political evolution. In establishing the principle of collective security and developing mechanisms, agencies and broader structures for its implementation, these states demonstrated their abilities to alter political expectations and authoritative structures. These features and sensitivities equip them to pursue effective international leadership in developing climate change mitigation and adaptation strategies.

The conditions, arrangements and expectations created by the principles and underlying ideas of collective security have influenced the negotiating structures and parameters within which global climate change is considered. They contribute to states' assessments of their relative risks, including their perceptions of trust and mistrust, friendship, alliances and enmity (Edmondson and Levy, 2008). The key sectors that influence states' policy- and decision-making are illustrated in Figure 3.1. In the context of global climate change, each of these sectors is subject to changing and sometimes uncertain expectations, making the guardianship tasks of government more difficult.

Existing international political structures and institutions accept global interdependencies, and yet states and international political structures remain premised upon high levels of state autonomy and independence (Paterson, 2000; Young, 1994). Although the latter half of the 20th century produced many instances of states willingly accepting

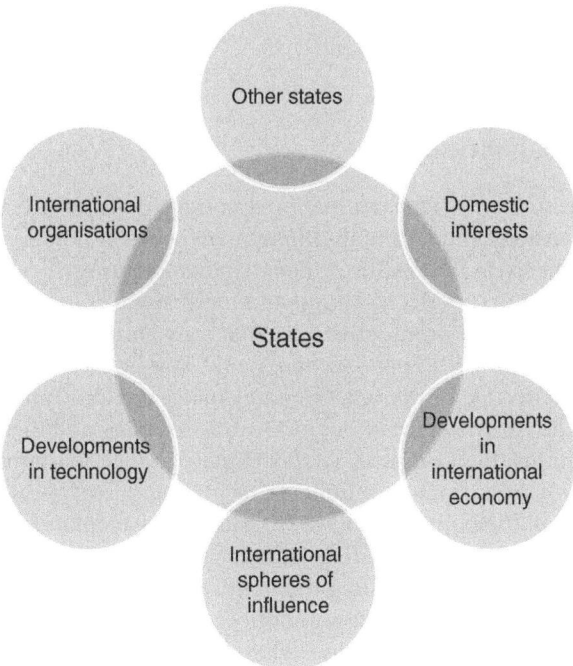

Figure 3.1 Sectors of influence over states' international policy choices

increasing levels of economic interdependence, there have been few instances where states have readily accepted complex collective political interdependencies. Changing views about the relationships between prosperity and progress are part of states' fluctuating expectations concerning their rights and responsibilities. They shape states' assessments of the potential risks of global climate change and decisions concerning how best to minimise these.

In the past, states identified and assessed their national interests in terms of their economic and strategic security, measuring their relative abilities to produce, trade, negotiate and coerce. However, since the 1990s, when international environmental problems began to attract more systematic attention, some states have preferred to adopt international institutional arrangements for managing and monitoring environmental problems (Biermann and Bauer, 2005; Young, 1994). Over the last couple of decades, it has become increasingly apparent that global climate change responses are challenging the nature of independent authority exerted by states as they experience new patterns of interdependence and new complexities in identifying and assessing their interests.

Reasserting political goals

Within the contemporary international political community, there are ongoing tensions and efforts to differentiate between politics and economics. Effectively addressing climate change requires the absorption of economics into politics to permit an orderly unravelling of the international political economic structures that were created through strong capitalist growth in the post–Second World War era. This is necessary because an effective response to climate change can only emerge from politically driven transformations of identity, societies, and the practices and aspirations they comprise. Global climate change solutions cannot and will not arise from economic or technological practices that are divorced from the political sphere. Mitigation and adaptation strategies must take full account of current and emerging production, trade and conservation patterns, including the political means by which they are supported, such as energy and water prices, subsidies and taxes (United Nations Development Programme, 2008; Hoffman and Hoffman, 2008; Kütting, 2000).

Politics entails shaping and managing the lifestyles and aspirations of human communities, which are essential to climate change responses and choices across the world. Economics and technology service the politically articulated preferences of societies. Effectively addressing global climate change therefore presents political challenges. Once appropriate policies and political structures are identified, economic and technological activities will need to be reshaped and coordinated to achieve their agreed goals (Roberts and Parks, 2007; Gupta, 2005; Kütting, 2000).

Interdependent responses

By and large, states continue to prefer international political arrangements, structures and institutions that consolidate their rights to political independence. Notwithstanding the growth of multilateral agreements and their implications for states' sovereign autonomy, states continue to assert and reassert their rights to political autonomy (Biermann and Bauer, 2005; Paterson, 2000). For instance, states continue to assert control over their borders and the flows of people and goods across them. They regulate the activities of key social and economic actors within their borders and they reiterate their supreme rights to determine their political structures and institutions.

Some leading contributors to global climate change discussions argue that the new economic challenges of food and energy production and water redistributions will drive states to unify their domestic and international political interests (Stern, 2009). We do not share their optimism. Indeed, we argue that contemporary states display ambivalence regarding their inherent physical, material, knowledge and decision-making interdependencies because they struggle to reconcile their divergent sources of political and economic interests. Without being unduly gloomy, we argue that economic challenges are likely to exacerbate political divisions and tensions, unless effective political leadership is exerted to prevent states from exerting ongoing independent decision-making. Such tensions are present in all international political engagements, even when these relate to more discrete and less complex areas of governmental decision- and policy-making than those associated with managing climate change impacts. However, in climate-related negotiations these dynamics increase states' reluctance to take even modest unilateral steps towards changed practices and visions, and reduce the likelihood of willing cooperation in multilateral forums.

Global climate change is an international political challenge, raising a package of issues that highlight the physical interdependencies among states, although many remain reluctant to respond to these. The actors who express interests in them reveal some of the ways in which global climate change presents new uncertainties concerning who is authorised to act, upon whose behalf, and who provides the means by which these actions might be taken. These political challenges also affect the circumstances under which political leadership might be demonstrated and the range of rewards and punishments that might pertain to those who meet or fail to meet previously agreed targets (Bearce et al., 2009; Postiglione, 2001) (Table 3.2).

Building enforcement and compliance is a basic problem for states and the international organisations they create to manage, monitor and implement their international agreements (Hempel, 1996). The vast majority of international environmental agreements permit only limited enforcement because the broader international context retains the primacy of independent sovereign states. Within international agreements, enforcement and compliance rely upon the positive influence of cooperation. Within international environmental agreements, specific sanctions, such as those that exclude parties from subsequent

Table 3.2 An overview of the political debates and challenges of global climate change

Cost-related debates	Prosperity-related debates	Progress-related debates
Implementation responsibilities	Prospects of economic prosperity	Political importance of economic growth
Establishing and maintaining implementing and monitoring agencies	Maintaining affluence and economic growth	Future implications of sustainable development
Costs of not meeting targets	Economically viable production	Distributions of economic rights
Distribution of costs	Sustainable development	Political values as drivers of human association
	Patterns of production	Supporting and managing changed forms of political organisation

negotiations, are rare. Instead, compliance relies upon states accepting roles as good international citizens and embarrassing recalcitrant parties by pointing out instances of poor conduct.

Liberal democratic paradoxes

Climate change is creating new paradoxes for liberal democracies, and for the political visions associated with them. On the one hand, it has exposed the international political community as requiring renewed attention to democratic and representative political practices since these will be required to enable fuller inclusion of diverse states – and also for spreading expectations about the value of preserving vulnerable societies. Ironically, liberal democratic political visions can provide more meaningful protections of vulnerable and marginalised peoples, because of their tendency towards greater inclusiveness of diversity and beliefs in individual rights. On the other hand, responding to global climate change also requires stronger levels of international governance, including strengthened mechanisms for enforcement and compliance (Paterson, 2000). As shown in Figure 3.2, states are subject to numerous sources of political interests and experience competing goals as they seek to make and implement laws and regulate the conduct of others.

Already, we are seeing an increase in bilateral initiatives towards monitoring the behaviour of neighbouring states, marking an extension of

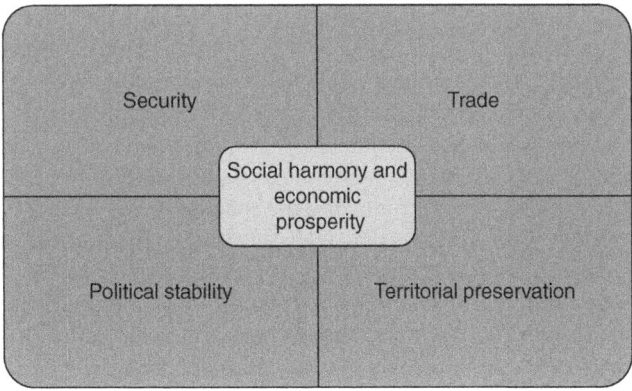

Figure 3.2 Key drivers of states' goals

states' efforts to ensure agreement compliance, especially when supporting domestic legislation is also required. In this way, states are imposing new levels of accountability upon each other (Vogler, 2005; Paterson, 2000). In so doing, they are manifesting their recognition of practical political equality within the international political community – an attribute often noticeably absent throughout earlier dealings wherein legal equality did not also produce equal political rights. Emerging carbon trading plans provide ready examples of this new dynamic in international climate change negotiations (Arup, 2009).

Conclusion

The environmental problems that became politically salient in the 1980s simultaneously challenged notions of national sovereignty and reinforced interdependent relations between states. They emphasised the interdependence of states and highlighted the biological and geophysical systems that underpin human and other life. The challenges of addressing increasing greenhouse gas emissions and ozone-depleting substances similarly present new challenges for the nature of sovereignty (Edmondson, 2011; Paterson, 2000; Mathews, 1991). While some of the specific and widely recognised impacts of climate change had not achieved wide public awareness by the late 1990s, the global environment was experiencing marked changes in the 'loss of species, ozone depletion' and large-scale 'deforestation' (Mathews, 1991, p. 31).

These pointed to new economic and political challenges that needed to be addressed by individual states as well as the international political community at large. These challenges required collective solutions because they entailed problems that 'render[ed] geographic borders irrelevant' and posed 'a major challenge to national sovereignty' (Mathews, 1991, p. 31). Early global climate change issues highlighted the interconnectedness of sovereign states and the reliance of governments upon knowledgeable policy-makers who were sensitive to the political dimensions of protecting shared resources (Barry and Eckersley, 2005; Young, 1994). It became clear that international environmental problems, such as global climate change, inevitably require collective action that can only be achieved through collective decisions and policies.

The key norms, practices and structures that comprise the international political community are in the process of being exposed as inadequate for the forms of international governance required for effective and sustained climate change responses. Global climate change poses new political challenges to achieve meaningful greenhouse gas reductions through multilateral conventions and protocols. The international political community confronts new challenges in responding to intensified issues concerning distributive justice and equity in terms of the costs of alleviating its worst consequences (Roberts and Parks, 2007; Gupta, 2005). Achieving effective, efficient and equitable climate change responses depends upon establishing new beliefs concerning the roles of governments, ensuring that states accept new responsibilities for the well-being of other citizens, and new capacities to exert authority over other social and economic actors (Roberts and Parks, 2007; Achterberg, 2001). As Hoffman (1997, p. ix) argues:

> what threatens us today...is...an imbalance between the supreme legitimate authority that still resides...in the sovereign state, and the incipient but fragmentary and feeble authority of collective institutions dealing with problems that transcend the states, or exceed their capacities, or require the reduction of their authority.

Increasingly, sovereignty relies upon international participation and accommodation of interests beyond those they might immediately identify as their own.

As global climate change alters political relationships and structures within the international political community, those states that do not respond to its causes and effects are likely to experience severe constraints upon their autonomous authority. States' future actions are likely to be guided by the actions of other states, intergovernmental organisations and non-state actors. The formation of an international environmental authority, such as a World Environmental Organization, is among the possible political changes that might arise in expanding international climate change governance (Biermann and Bauer, 2005). The international political community has already demonstrated its capacity to establish and impose enforcement measures such as sanctions or multilateral intervention to compel compliance with norms and modes of behaviour outside the autonomous decisions of individual states (Kütting, 2000). Many of these provisions presently relate to preventing and punishing abuses of human rights but similar mechanisms might be applied to states that fail to act with regard for common environmental interests. Such multilateral actions would override historically established norms of non-intervention and overturn outdated notions of absolute autonomy.

Effective international responses to global climate change and its unequal impacts will require systematic recognition of the sources of divergent interests and governmental capacities. New sources of international political influence now impel states and other key international actors to take account of competing and conflicting interests and priorities beyond themselves. Their presence and the issues they raise for public debate reveal doubts that earlier patterns of progress can be sustained through continued reliance upon state-based sovereign authority. For instance, increased life expectancies, greater production and enhanced life opportunities, which were hallmarks of progress through industrialised production, associated technological developments and economic progress, can no longer be taken for granted as future prospects.

4
A Rowdy and Unruly Community

Introduction

The international political community confronts a significant difficulty in addressing global climate change. Quite aside from the contested nature of the science that underpins the anticipated consequences, there are pressing international political problems that demand resolution. The canvas of voices seeking to express concerns or to participate in debates concerning the nature and impacts of global climate change incorporates a broad spectrum of social and political positions from environmentalists, scientists, economists, philosophers and industrialists. Their suggested responses to global climate change range from radical social, political and economic reform, such as the wholesale overturning of modern industrial capitalism, to advocating renewed faith in progressive technologies. Among them are appeals to fundamentally re-imagine human relationships with nature and property that would lead to reshaped human lifestyles and aspirations (Eckersley, 2005).

Some approach the questions of managing global climate change by seeking to reform key political values first and the forms of association they create second (Conca, 2005; Dryzek and Schlossberg, 2005; Dryzek, 1997). Others suggest that faith should be placed in current forms of technological development, arguing that '[w]hat is needed' to enable effective responses to global climate change impacts 'is faster progress in a direction in which we are already headed' (Pearce, 2007, p. 247). Others point out that this direction is destructive and likely to exhaust the world's fossil fuel–related resources within the near future, leaving human populations vulnerable to resource shortages and poorly equipped to deal with changing climatic patterns and the weather patterns they produce (Newman et al., 2009; Holdren, 2006; Brown, 2005;

Anderson and Leal, 1998). These views are hallmarks of the rowdy and unruly nature of the expanded array of voices that comprise the international community in the 21st century, marking its shift from relative order and stability in diplomatic, trade and collective security arrangements that characterised the last decades of the 20th century.

In the midst of this cacophony, private actors, social networks, economic corporations and political actors of various kinds seek to influence the scope of debates and to move leaders towards specific decisions because they perceive global climate change as driving changed political and economic practices and forms of association (Archibugi and Held, 1995; Pogge, 1992). Many advocate urgent action, arguing that the 'future' is now, and the world's ability to support an increasing population is already severely compromised (Jackson and Sorensen, 2007; Cox and Jacobson, 1997; The Commission on Global Governance, 1995). Across this spectrum of actors, common positions have emerged, with sometimes unexpected sources of cooperation, such as between economists, scientists and business leaders. For instance, some of the world's largest companies have expressed support for dramatic reductions in carbon emissions of between 50 and 80 per cent by 2050 (Kanellos, 2008; Van Noorden, 2008).

This array of contention suggests an emerging recognition that viable, achievable and widely supported responses to climate change will be, first and foremost, a political achievement. However, there are also continued anxieties about the extent and distribution of the economic costs entailed in implementing such measures, which might 'wreck our economies and put the lives of tens of millions of people at risk over the coming decades' (Dimas, 2007). These anxieties and ongoing inequalities in political and economic capacities limit the formation of coalitions of interests among political leaders and contribute to delays in global mitigation and adaptation strategies. As the level of understanding of the impacts of global climate change increases, and the range of actors who seek to respond to such knowledge expands, pockets of consensus will provide a necessary platform for political action (Edmondson, 2011).

Accommodating these diverse voices in responding to global climate change creates pressures for an international political community that is structured according to hierarchies of power arising from the material capacities of sovereign states. According to established international political practices, maintaining territorial security overrides other issues which increase the political challenges of responding to global climate change (Barnett, 2001; Dyer, 1996). Disciplined, coordinated and achievable climate change responses can only emerge through political processes. This does not mean that new and potential technologies will be excluded from diverse efforts to maintain modern industrialised economies, lifestyles and their forms of political association (Hawken, 2005). However, continuing to believe that science and technology will solve climate change problems amounts to wishful thinking motivated by unwillingness to confront the lifestyle sacrifices required of people in all societies.

In spite of accumulating knowledge and increasing experiences of climate change impacts, the transformative effects of global climate change are yet to be fully appreciated. Two factors account for this. The first stems from uncertainty and ongoing debate about the accuracy of the underpinning science. The second stems from uncertainty about the industrial and political changes required by climate change responses. Both of these contexts of debate have been shaped by the dominance of liberal-progressive political values and ideas within the international political community.

An expanding range of voices

The 20th-century liberal project of encouraging and extending democratic practices expanded the range of voices that now claim rights to participate in domestic and international political sectors (Vogler, 2005; Holden, 2000). It is becoming increasingly evident that the political effects of global climate change will result in a world that exhibits remarkable differences from present forms of political organisation. Although the features of this changed world remain indistinct, it is likely to involve the end of the 20th-century liberal project of extending democratic processes across the globe. The international community is unlikely to be able to extend the liberal project of spreading democracy by simultaneously supporting ongoing quests for economic progress, addressing global climate change and supporting efforts to adapt to the altered political circumstances it creates. Even before widespread challenges to economic progress have been experienced, some important contributors to climate change debates argue that democratic and industrialised societies appear to be a flawed basis of future development (Heyd and Brooks, 2009).

Accommodating the diverse array of new voices that claim rights to participate in developing strategies to address global climate change while states adapt to new climate conditions will create a political environment with less restrictive membership than currently exists.

At present, states hold pre-eminent authoritative status and their authority structures form the basis of international decision-making (Edmondson and Levy, 2008; Conca, 2005; Rabkin, 2005). Currently, states can override non-governmental organisations (NGOs) and inter-governmental organisations (IGOs) because they rely upon the specific forms of authority delegated to them (Higgott et al., 2000). Accommodating new voices to minimise dissent and maximise cooperative responses to global climate change will be necessary to maintain and promote order. This presents a political challenge since conventional wisdom concerning effective international agreements suggest they rely upon shared or common perceptions of problems and shared views of appropriate solutions (Kütting, 2000; Young, 1994). As the number of interested actors increases, it becomes more difficult to accommodate their diverse interests.

There are long lead times in achieving agreements to enable international strategies. Even after such agreements have been reached, their implementation entails complex processes that involve protracted timeframes. The Kyoto Protocol, as a pertinent example, took nearly eight years to enter into force, although doing so required ratification by only 55 states (United Nations Framework Convention on Climate Change, 2011). Global climate change impacts are already occurring and will continue for long periods even if their causes are promptly addressed. Just as preserving a habitable environment for future generations is an imperative for current leaders and thinkers, so too is planning the preservation of a desirable political system.

Achieving effective international responses to global climate change relies upon widespread acceptance of the costs to development and modern consumption-based lifestyles within a framework of a morally defensible and broadly accepted distribution of these costs (Page, 2006). These require states to identify broadly accepted forms of political association and structures of authority. They also require collective actions by states and international institutions to achieve internationally coordinated protocols and instruments to harness the activities of diverse actors who are committed to addressing global climate change (Vogler, 2005; Haas et al., 1993). Their success depends upon accepted instruments and processes to compel states and other actors to comply with agreed targets and other initiatives arising from international agreements.

A universal acceptance of the need to curb ongoing industrial development and associated aspirations of affluence will be required to maintain international order and address global climate change. Only in this way will it be possible to preserve a natural environment that is capable of sustaining an expanding global human population. To achieve this, there needs to be widespread acceptance of the science underpinning the case for global climate change and the steps necessary to address it. While the former appears to be ever closer, such gains are meaningless without the latter. Achieving the latter requires the fuller accommodation of the diverse voices that clamour for attention in this new political context.

At this stage of the 21st century, governments and people must begin to come to terms with the likelihood that timely and appropriate outcomes are doubtful. Existing institutions have quite poor records in transcending or otherwise depoliticising efforts to sustain international peace and development, and even worse records in promoting redistributive justice (O'Neill, 2009; Page, 2006). Nonetheless, these are the tasks that confront the international political community as global climate change impacts unfold. It is highly probable that maintaining order, democratic principles and accepted notions of progress will prove extraordinarily difficult as political, economic and social actors are compelled to address global climate change.

A very rowdy community

In the late 20th century, the rise of 'new voices' desiring political participation within states and across the international political community influenced the nature of states as sovereign actors. Their accommodation and recognition in many parts of the world increased the diversity of political interests influencing domestic and international policy-making. By the end of the 20th century, a plethora of non-state actors had become accepted collaborators in international forums after gaining fuller voices in international politics at the 1996 Rio Earth Summit. At this important early international climate-related summit, states were joined by 18,000 NGOs, running a parallel meeting, and observer parties who were permitted partial participation in discussions and negotiations (Friedman et al., 2005, p. 36). Subsequent international conferences and policy-making forums have continued to extend the public voices of non-state actors, providing them with opportunities to seek more direct forms of political influence. Hence, in the 21st century, NGOs play important roles in international agenda setting and help to focus political debates even though they do not hold formal status or voting rights at international conventions.

These new voices require the international political community to take fuller account of complex and challenging political issues concerning rights and responsibilities in decision-making. Many of these voices in the international political community are likely to matter for responses to climate change because they have created new opportunities for international cooperation and expanded the field of interests requiring accommodation. By acquiring political influence, they have created new and unresolved tensions concerning the nature of political authority, legitimate sites of political identity and international decision-making. Their presence has increased the complexities experienced by contemporary states in identifying their national interests, further blurring decision-making and authority boundaries. These new actors, and the negotiating sites they produce, create new opportunities for cooperation and expand the range of interests and views of possible and preferable outcomes.

Many of these new voices have emerged from agencies of transnational capital, international aid, and development (O'Neill, 2009, p. 15). Many, such as the United Nations (UN), the World Trade Organization (WTO) and the IPCC are products of the industrial and technological 'progress' of the 20th century and the forms of citizenship promoted by it. Their politically interested voices have influenced state-based and international political processes. They have formed advocacy and other interest groups.

Previously overlooked groups have gained voices and currency in the international political community, creating a denser and broader web of global relations. They have contributed to the emergence of formal and informal policy networks and renewed international debates concerning the rights of states and their individual and collective capacities. Non-state actors now contribute to states' interests and identities (O'Neill, 2009; Eckersley, 2005; Laslett, 2001). Some sub-state actors, such as Greenpeace and the World Wildlife Fund, have become international actors in their own right informing and shaping the interests of states and intergovernmental agencies. Key economic actors, including European oil companies, have also become engaged more fully in public debates concerning international and domestic climate change policies seeking to increase their participation in the political processes surrounding global climate change responses (O'Neill, 2009, p. 63).

New sources of international political influence, such as those outlined in Table 4.1, encourage states and other key international actors to take account of competing and conflicting interests and priorities beyond themselves. Their presence, and the issues they raise for public debate, confirm doubts that earlier patterns of progress cannot be sustained through continued reliance upon state-based sovereign

Table 4.1 Selected examples of new voices

Key non-state voices in climate change debates	Examples
NGOs	World Wildlife Fund, Greenpeace
IGOs	United Nations Environment Programme, World Trade Organization, United Nations Framework Convention on Climate Change
Multi-state associations	EU, low-lying and island states (AOSIS)
Key economic actors	Electricity producers, motor vehicles manufacturers, oil producers
People as social and economic units of states	Green movements, adopters of water restrictions, lobbyists, purchasers
Scientific and knowledge and communities	IPCC, professional bodies examples, Scientific Committee on Problems of the Environment (SCOPE)

authority (United Nations Development Programme, 2008; Conca, 2005).

Interdependent interests and authorities

Understanding how the international community is responding to climate change requires sensitivity to the diverse processes through which solutions are being sought. Throughout the 20th-century historically based understandings of states as unified actors were reinforced and their actions understood as consequences of national interests, which were defined by broader internal imperatives of self-promotion. This reinforced a dichotomy between the 'inside' and 'outside' of states with a corresponding 'ours' and 'theirs' evaluation of interests and identity (Doty, 1996) (see Chapter 11 for a fuller discussion of these issues). These dynamics supported views of states as unique, progress-driven authoritative entities.

For several centuries, states fiercely protected their autonomy, arguing that this was a key right arising from their status as sovereign authorities (Rabkin, 2005; Bonanate, 1995). States were concerned with their material capabilities as sources of political independence which diverted attention from other ethical and dispositional considerations. However, in the latter stages of the 20th century, the nature of states' autonomy came under question as doubts emerged about the enduring utility of 'classically' independent states (Oye, 2005; Ohmae, 1995). Human rights and environmental debates, in particular, constituted sites of contested rights and responsibilities among and between states where new forms of authority appeared desirable.

A universalised belief that states held unified interests within their borders perpetuated their claims to exclusive control over the use of resources within their territories (Edmondson and Levy, 2008; Eckersley, 2005; Anderson and Leal, 1998; Dryzek, 1997). It also provided a central foundation of the international political community, enabling states to develop their economies and exert political authority that consolidated sovereign independence. Such a historically based view now suffers from its inability to acknowledge the expanded array of political processes and demands that occur within states. It also discounts the linkages between states in their international dealings, as well as among non-state actors and organisations. In recent decades, these historically based views have been challenged by non-state actors claiming forms of legitimate authority. Although state-based forms of authority and political association remain dominant, the rowdy voices of diverse interests are ensuring that sovereignty evolves in response to new social forces and political imperatives (Conca, 2005; Dryzek and Schlossberg, 2005; Philpott, 1997).

States ultimately retain their authority and primacy as political actors because, at the end of the day, only they remain accountable for the conduct and welfare of their citizens (Edmondson and Levy, 2008). For democratic states, especially, this includes providing opportunities for divergent voices to be heard from across the spectrum of interests upon which debated issues, such as global climate change, impact (Hunold and Dryzek, 2005). Their need to accommodate and respond to the influential presence of such entities can also contribute to poor and untimely decision-making as they become susceptible to pleasing and appeasing interest groups.

Accommodating diversity

Climate change has become a key issue in global politics as the rowdy voices of the international political community have drawn attention to the threat climate change poses for order, states' sovereignty and their rights of absolute independence. As international political debates and negotiations have shifted focus from security and the economy to include environmental and human rights issues, more groups have achieved credible political presences (Friedman et al., 2005). Accommodating diverse non-state actors in climate change debates extends states' recognition of complex political contexts in reconciling their claims to exert legitimate authority over resources, territory and citizens and their rights to represent their citizens' multiple interests in international political forums (Achterberg, 2001; Holden, 2000). These challenging dynamics are particularly pertinent to democratic states where freedom of speech and rights to protest form cornerstones of representation and encourage active public participation by a diverse range of political voices.

To maintain legitimate authoritative status, within an increasingly environmentally conscious international community, states' representatives will feel increasingly pressured from within and without to implement foreign and domestic policies that mediate the effects of climate change. Inaction or perceived inaction, without extensive and time-consuming efforts to justify and rationalise both domestically and internationally, will draw condemnation from their own (or foreign) citizens. This democratisation of international debate about climate change is a testament to the success of the 20th-century liberal project and the corresponding representation of states' decisions. As global climate change impacts unfold, non-state and global media actors increasingly assume agenda-setting roles in seeking to influence states and their decisions (Sprain et al., 2009). Media and other non-state actors contribute to the validity of states' actions, influencing and critiquing their policies and giving voice to the agendas of non-state groups (Gilboa, 2005; Manheim, 1984).

It is no longer possible to ignore different voices as the expansion of global media agencies offer non-state actors (ostensibly) politically independent outlets through which to raise awareness of their various agendas. While media is not as ubiquitous as some suggest, communication technologies have contributed to the proliferation of interconnected social, political and economic networks and organisations. Myriad actors use communication technologies such as the internet and more traditional forms of media to highlight important issues, acting as political actors in their own right by influencing agenda-setting activities and providing opportunities for exerting influence by sub-state actors (Minion et al., 2009; Gilboa, 2005; Manheim, 1984).

Institutions and leadership

As shown in Table 4.2, many internal forces are now pulling political action in divergent directions beneath the surface of 'state' institutions and structures. In this context, state identities are neither singular nor unified. This adds another dimension to the complicated processes by

Table 4.2 How new voices challenge assumptions about states

Voices	Assumptions	Sources of political authority
Only sovereign states	Primacy of sovereign authority and secure borders	Independent authority Treaties
Sovereign states and IGOs)	Primacy of sovereign authority through legal equality supported by sovereign legitimacy and moderated by international law	Modern international law Collective security
Sovereign states, IGOs and NGOs	Collective interests impact upon individual state interests, contesting authority across more porous borders	Introduction of non-state interest groups into international politics
Sovereign states, IGOs, NGOs and international civil society	Evolutionary phase of sovereignty characterised by accommodation of non-state actors	Dramatically expanded range of voices, actors and issues

which state leaders seek to determine the nature and scope of their national interests and how best to secure them. In turn, states' external identities are shaped by their responses to debates and conflicts occurring within their borders.

Further complexities have been added to the international political community by communication technologies that advance the voices of media outlets and bring events into the living rooms of people around the globe. Interconnected media networks allow non-state and trans-state actors to express their views more widely and with greater levels of persistence. A political consequence of such international interconnectedness is to encourage those states that are already accountable to their domestic electoral systems to become accountable to an additional range of external interests. In contemporary international politics, NGOs and other non-state issue-based groups are achieving global prominence as they, and their audiences, are no longer contained within borders.

Power and identity

Whether and how carefully states choose to listen to the voices of groups within their borders depends upon their political systems. Liberal democracies that are concerned with representing the interests of citizens are only part of the context for new voices in international relations. With increasing global 'democratisation', many voices that were previously suppressed in autocratic or authoritarian states have attained greater prominence, both domestically and on the world stage (Achterberg, 2001). This global proliferation of international non-state actors has increased the complexity of the international political community and highlighted the plural identities and interests of states.

Different non-state identities, and the voices they have developed, highlight the limited authority and capacities of states in responding to existing global climate change debates, even though some of these represent views rejected by the mainstream politics of states. Non-state actors can also highlight minor or hidden issues previously ignored by governments, thereby increasing the risks that some governments will appear negligent. Liberal-democratic states will not deny these voices. This would require stringent media regulation, which contradicts freedom of expression. By not restricting these voices, it becomes increasingly impractical to deny their presence as agents of political reform. By denying different voices, however, states neglect valuable resources of ideas and information.

Analysts of politico-economic relations between states face predicaments in determining the different levels of power and influence exerted by non-state actors who now engage in political lobbying and seek to influence government endorsed think tanks. Identifying which voices exert greater influence within states and across the international political community are important aspects of their considerations. Determining which non-state actors carry greater legitimacy and validity, and justifying their recognition, poses additional political challenges.

States do not limit their authority by accepting the constraints that follow from membership of international organisations and adhering to treaty law and norms (Conca, 2005; Haas et al., 1993; Kratochwil, 1989). Their decisions are always 'expression[s] of choice' that reflect their authority since international recognition and legitimacy are expressed in international activities (Rudolph, 2005). States have moral obligations to enforce environmentally responsible practices to improve cooperation with citizens, industry, corporations, international organisations and non-state actors (Markowitz and Shariff, 2012a; Low and Gleeson, 1998; Dryzek, 1997). Ignoring or neglecting global climate change is costly for states because it limits their participation in international decision-making and reduces their regulatory capacities.

Order and cooperation

In the 21st century, international order relies upon widespread acceptance of global climate change responses that restrict states from claiming privileged autonomy. Whereas states once held supreme rights to exercise authority without intervention from external parties, their independence is compromised by shared reliance upon global commons resources and new imperatives of climate change mitigation and adaptation strategies (Brauch et al., 2008; Barnett, 2001). These new dynamics hold priority over the political expediency of non-intervention among states. In the future, breaches of climate change agreements could potentially justify intervention as the international political community reacts to the political and security challenges it presents. Global climate change will alter perceptions of international security such that the practices and frameworks of collective security are expanded towards responsibilities based forms of political association (Brauch et al., 2008) (see Chapters 5 and 10 for elaboration of conflict-related issues).

Responding to global climate change requires cooperation from all states because these effects will be felt by all and the required sacrifices to independence, economic prosperity and social well-being should not be assumed by only a few. Expecting some communities to assume unfair burdens of sacrifice will create new tensions between those prepared to act in the interests of the global community and those myopically concerned with their own self-interests (Page, 2006; Najam, 2005b). All states need to acknowledge the severity of the problems presented by global climate change and the magnitude of generating solutions by assuming responsibility for developing appropriate policies and regulatory environments. The presence of international order and the institutions arising from it, such as diplomacy and international law, suggest it is plausible to believe the international political community is capable of developing solutions to which the majority can contribute. The timeliness of effective responses, however, remains an issue of genuine concern.

Scientists and communities around the world are becoming increasingly aware of the extent to which an interconnected ecosystem underpins the humanly constructed units of differentiation known as states (Hughes, 2009; Huggett, 2006). Curbing the effects of climate change will require cooperation between states and non-state actors including industry sectors as industrialised states seek to regulate the emission of carbon from the activities of private industrial actors. Such measures will ultimately extend beyond state borders to include those activities that are routinely outsourced to market-based suppliers in developing states (Conca, 2005). These decisions and their implementation depend upon the cooperation of economic actors including multinational corporations that are accountable for adverse contributions to climate change, the regulation of which is more difficult as the globalised nature of large businesses collides with the localisation of state-based business law and practices.

Non-state actors increase the complexity of maintaining international order. Ideas about global security will need to be expanded to accommodate these new actors and to provide them with a level of legitimacy (Baylis, 2008; Karmad, 2008). Granting political status to an expanded array of actors might support global security by co-opting the energies and expertise of new political voices to ensure they become part of the solution. Regulating and considering actors other than states, or paying attention to the ways in which states are not simply unitary actors, dramatically complicates global security and order. The need to recognise a range of alternative actors and establish their legitimate participation is likely to become a reality for maintaining order in the 21st century.

Addressing the problems of climate change and water redistribution requires the international political community to declare the primacy of sovereign states outdated. This does not mean states will cease to exist as some, such as Ohmae (1991), have previously advocated. Rather the nature of states' authority will evolve to accommodate other voices. Actors such as environmental groups, scientific communities and industry associations are becoming increasingly important because their voices can no longer be denied. Democracy has created political spaces for their participation at both the domestic and international level. They have claimed rights of participation alongside states and altered the nature of sovereign authority to secure their roles as legitimate sources of political interest.

Conclusion

Retaining unquestioned reverence for sovereign authority in its traditional form is unlikely to enable effective and timely responses to climate change impacts, including water and resource redistributions and conflict. Maintaining geographically prescribed interests and issues, which have become politicised through long histories of competition, is unlikely to encourage timely responses to global threats. Accommodating non-state voices as legitimate political actors is likely to aid the development of shared responsibilities that transform the international political community by diminishing the influence and predominance of state-based interests. However, the participation of non-state actors within the international community will threaten the traditionally conceived sovereignty of states. Currently, it is sovereignty that provides states with authority to confer legitimacy to non-state actors. To accommodate new voices, the nature of sovereignty will need to further evolve, allowing action by an international political community composed of a variety of actors with varying sources of provisional authority.

Expanding the capacities of economic actors, environmental lobbyists and groups with scientific expertise to influence international political arrangements may improve order and security in the 21st century. Accommodating these diverse actors can strengthen ideas, knowledge and information flows among states and aid in reducing confusion and uncertainty. If conventional thinking is correct, it is uncertainty that leads states to experience an ongoing security dilemma characterised by distrust, fear and conflict (Buzan, 1991). In the 21st century, states cannot afford to be perpetually self-interested or to consider themselves independent units because the unevenly distributed global effects of climate change will ultimately affect everyone. The most effective means of addressing climate change challenges lie in the participation of states and citizens in developing, promoting and contributing to solutions.

Maintaining order requires shared knowledge to counter uncertainty that gives rise to distrust, fear, conflict and war. Recognising the contributions of knowledgeable voices may provide a more robust international political community equipped to address the challenges of global climate change in an orderly manner. For this reason, membership of the international political community must be expanded to accommodate actors with relevant expertise and knowledge to support states in addressing global climate change. Stifling debate by recognising only state-based interests will not assure timely or universally acceptable climate change responses. Channelling non-state interests and expertise through state-based political processes will also curtail effective global responses.

5
Water, Disorder and Disrupted Development

Introduction

Many states will find themselves caught between a proverbial rock and a hard place as they seek to develop policies and regulations to meet the need for secure energy sources and securing food supplies in a changing global climate (Zhao and Running, 2010; Battisti and Naylor, 2009). Government planning cannot prevent less predictable rainfalls, more severe storms, droughts and floods, and yet governments will ultimately bear responsibility for their repercussions. Worldwide, it is governments that oversee and enable the development and maintenance of infrastructure to support modern societies and the industrial activities that sustain them. In the 21st century, industrialised states will grapple with an expanded array of challenges in ensuring the wellbeing of their people, economies and social sectors. Climate change consequences will routinely present 'predictable surprises', upsetting budget projections, demanding emergency responses and new forms of policy coordination (Bazerman, 2006; Bazerman and Watkins, 2004). These challenges will be even greater for states that lack established and adaptable infrastructure and governmental capacity (United Nations Development Programme, 2008).

As attempts are made to reduce carbon emissions, states will experience political and economic difficulties because their decisions impact upon production, prosperity and lifestyles. For instance, as some governments work to make 'the use of carbon sources more costly' their actions will have direct impacts upon their citizens and corporations (United Nations Committee for Development Policy, 2009, p. 9). At least, in the short term, wealthy citizens in affluent states may well be able to afford to pay more for the water and electricity they consume in their homes and recreation centres, while poorer members of societies will not. The assumption that the costs of climate change can be borne makes mitigation policies more contentious even when they are recognised as necessary. Such approaches imply that issues of environmental sustainability and resource security are matters of economic capacities and ignore the scientific evidence and predictions. Although wealthy citizens in affluent locations might be able to pay more for the water they consume, they will not be able to produce higher rainfalls and they are unlikely to be able (or willing) to simultaneously pay more for water, electricity, transport and food.

Introducing 'affordability' as a measure for determining the suitability of climate change mitigation strategies raises a divisive and false dichotomy in political debates. As observed by O'Brien and Leichenko (2000, p. 225), '[e]xtreme climate events can impact the wealthy and poor alike, particularly in high risk environments': while wealth may provide a measure of resilience it does not reduce vulnerability. Affluent states might be better able to afford to install booms across rivers to reduce up-river sea-water flow, but many of these states also rely upon established infrastructure for water catchment and electricity production and distribution. The fixed nature of such infrastructure, combined with extensive costs and long lead times in replacing or adapting production practices and developing new technologies reduces their flexibility in responding to these challenges.

Overall, such arguments provide rationalisations to forestall collective action and overlook personal impacts that are equally disruptive to the lives of individuals. Ultimately, governments of industrialised states will need to impose carbon and other greenhouse gas emissions limits, while simultaneously building new infrastructure to support the implementation of new technologies. These are the necessary steps required to mitigate the direst predicted climate change consequences and to transform real and aspirational lifestyles into more environmentally sustainable patterns. Confronted with these requirements, even the most prosperous of states will struggle to maintain buoyant economies, and democracies will struggle to sustain electoral mandates for climate action.

Climate change cannot be addressed by tinkering at the edges. Instead, it requires fundamental reforms to development, consumption, lifestyle practices and expectations. One way or another, 'increased oil prices, the application of a carbon tax, or dealings within emissions markets' limit economic prosperity even as they positively contribute to increased sustainability in societies (United Nations Committee for

Development Policy, 2009, p. 9). In terms of economic impact, policies that seek to limit rates of energy consumption and other inputs to production, such as water are 'regressive...and will ultimately halt development' (United Nations Committee for Development Policy, 2009, p. 9). Consequently, political leaders face difficult choices as they seek to preserve their societies and adjust to changing environments.

It is politically and economically problematic to imagine limiting or halting progress in societies that are already industrialised. Doing so in developing states is even more difficult to justify. The vulnerabilities arising from poverty, poor infrastructure, uncertain food and water supplies often coincide with governments that hold very limited means of growing their economies while sustaining political and social harmony (United Nations Development Programme, 2008; Roberts and Parks, 2007). The additional burdens imposed by initiating and adopting climate change mitigation policies and practices compound what are already politically demanding circumstances in many states.

The water-related impacts of climate change, such as changed rainfalls and rising sea levels are likely to jeopardise many efforts to promote development and prosperity among all states (Holgate, 2007; Huntington, 2006). Developing states in Africa are becoming vulnerable to dangerous climate change impacts, especially as surface water levels diminish (United Nations Development Programme, 2008; de Wit and Stankiewicz, 2006). Many populous cities and regions will be affected by storm surges and rising sea levels with 3351 cities and 10 per cent of the global population situated on the vulnerable low-elevation coastal regions (United Nations Human Settlements Programme, 2008). While developed states face economic and political challenges in responding to global climate change, they might be expected to have levels of resilience that might now enable their timely adaptation of habitation patterns, infrastructure and consumption of resources.

By contrast, developing states are poorly equipped for the magnitude of costs associated with the effects of climate change upon rainfall, food production, disease and increasing water and soil salinity levels (Oxfam, 2007). Even without the added burdens of climate change, these vulnerable societies are poorly equipped to respond to changing economic and social conditions. Climate change presents multiple tipping points that render them vulnerable to political and social upheaval, economic collapse and repeated natural disasters. These realities are exacerbated by population growth in regions that will experience significant water shortages (United Nations Educational, Scientific Cultural Organization, 2009). As Postel (2000, p. 941) observes, '[w]ith the world population projected to increase by an additional two billion...by the year 2030...finding ways to satisfy humanity's water demands ... now ranks among the most critical and difficult challenges' confronting the contemporary world.

The expected redistributions of water resources across the world present major challenges to the economic, social and political security of many people and their governments and challenge the mechanisms, structures and organisations that currently attempt to ensure international order (United Nations Educational, Scientific and Cultural Organization, 2009; Jackson et al., 2001; Postel, 2000). The uneven impacts of water-related climate changes contribute to confusion among states and other actors. Seasonal variations and emerging new climatic cycles increase the difficulties confronted by political decision-makers (Huntington, 2006). Currently, water policies are often contextualised at local, national and regional levels, but climate change requires globally coordinated responses that necessitate new international legal frameworks for water management, consumption and use (United Nations Educational, Scientific and Cultural Organization, 2009). Such steps are paramount if the world's finite and shifting water resources are to be ethically and effectively managed in order to minimise disorder and disrupted development.

In spite of myriad political complexities and administrative challenges, many governments have begun to act in setting emissions targets and supporting the commercial availability of alternative energy sources (United Nations Development Programme, 2008; Oberthür and Ott, 1999). In recent years, the costs of inaction have become clearer and scientific analysis has shifted beyond measuring temperatures and monitoring water flows, species loss, rising tides and desertification to considering the environmental burdens of maintaining present social systems (Wackernagel et al., 2002; Vitousek et al., 1997). Affluent states have begun to assess their vulnerabilities by measuring the costs and capacities of sustaining cities while also producing and distributing food for an expanding population (United Nations Human Settlements Programme, 2008). As has already been noted, global climate change presents additional problems for developing states where many populations are already struggling to eke out basic existences.

Agricultural economics

In the 21st century, as the planet's climate changes, many of the agricultural practices that were successful throughout the 19th and 20th centuries will need to change. These requirements give rise to political challenges in diminished prospects of progress, prosperity and democracy. In the 20th century, a universalised, one-size-fits-all approach to agricultural science enabled well-documented benefits in increased yields of key crops and increased food security. However, the production practices that brought these benefits also created multiple problems, including species loss, monoculture crop dependencies and environmental degradation (International Food Policy Research Institute, 2002). Although the green revolution of 1960–2000 produced notable successes in terms of agricultural production, these were achieved in the context of a particular set of climate conditions and were uneven across crop types and regions (Evenson and Gollin, 2003). The magnitude of human reliance upon our natural environment is illustrated by the United Nations' anticipation that global food production will need to rise by 50 per cent by 2030 (Owen, 2008). A changing global climate challenges further agricultural innovation (Tilman et al., 2002).

Agricultural activities account for 70 per cent of fresh water use and are susceptible to negative climate change impacts on the hydrological cycle (United Nations Educational, Scientific and Cultural Organization, 2009). For the populations of developing states who are directly reliant upon agriculture the importance of addressing these extends beyond food security (International Food Policy Research Institute, 2009). Their experiences of hunger, poverty, conflicts over resource security and patterns of refugee migration create 'political and economic disruptions, and social unrest' and put international stability and order at risk (Oxfam, 2007, pp. 9–10). Agricultural responses to climate change require improved global data concerning climate change effects and adaptation in agricultural practices (International Food Policy Research Institute, 2009). Political support is necessary to develop new policies and programs, increase investment in research, productivity and programmes for information and technology sharing.

Climate change has the potential to disorder and disrupt the stability and development of states and therefore requires fundamental reconsiderations of the relationships between political and economic spheres. These include those relationships at the heart of international political structures and the increasing numbers of rowdy voices clamouring to be accommodated in international decision-making (see Chapter 4). These dynamics present a challenging need to reverse the order of policy priorities between the economic and political spheres. At least since the middle of the 20th century, the principal considerations of governments have focused upon building their economic capacities, giving free rein to markets in some contexts and regulating them in others, in relentless pursuit of economic growth and prosperity. The contemporary international political economy acquired its institutional form in the same time frame, and this will also require substantial modification.

Unsustainable economics

Many of the obstacles that have created disappointingly patchy and slow steps towards global climate change responses lie in these political challenges. The unruly and untamed nature of global capitalism has shown itself to be more concerned with production, consumption and competitive advantage than the welfare and security of communities. Responding to the myriad implications of global climate change makes clear that responsibilities for common well-being ultimately lie with states as primary sites of political authority rather than facets of the international economy. Confronting and addressing global climate change in the 21st century requires an altered international political community in which politics drives economics as production practices, patterns of habitation and political associations experience upheaval.

Supporting those most vulnerable to the near-future impacts of climate change has encountered further political obstacles arising from contested claims to prosperity and progress. As discussed in Chapter 4, the competing interests of diverse actors and sectors create political impasses that states and the international political community are often poorly equipped to resolve. In recent years, these complexities have been exacerbated by disputes concerning the distribution of rights and responsibilities among states and others (Dryzek and Schlossberg, 2005; Eckersley, 2005; Thompson, 2001). Many of these concern the relative merits of regulatory mechanisms and the uncertain capacities of authorities to act logically and decisively. Ongoing debates concerning the merits of carbon cap and trade schemes, versus hard emissions targets, and direct taxes or water 'use and capture' licensing schemes provide relevant examples. The advent, adoption and cynical manipulation of slogans such as corporate social responsibility and green-washing provide ample evidence that unregulated economic activity remains exploitative and lacks an ethical concern for the welfare of the world's peoples above the pursuit of profits. Further complexities also arise from the international political economy and the broader context of security within the international political community.

The international political economy is a construction that serves to maximise the efficiencies of a global market by bringing producers and consumers into close contact and tight economic dependencies (Adger, 2006, 2000). In doing so it has contributed to global stability by incorporating peoples, states and resources into a single-functioning market system in which disturbances and disruptions brought on by conflict are undesirable. While the advantages and disadvantages of the international political economy might be debated along ideological lines, the underlying source of its success has been affordable bulk transport fuelled by pollution producing fossil fuels (Pearce, 2009). In a future when such fuels will not be plentiful, cheap or climatically responsible, the foundations of the international political economy may become less sustainable, providing likely sites for conflict between states and other political and economic actors (Stalley, 2003).

Market flaws

A range of serious consequences for producers and consumers who have become accustomed to a range of global markets and products are now emerging. In the short term, it is easy to foresee rising product costs as transport becomes more expensive as a consequence of carbon permits, rising energy/fuel tariffs or energy shortages. In the medium term, global markets will contract as the distance between sources and consumers again become significant and local products become increasingly price competitive. Additional market complexities are embedded within these patterns and economic practices. As Mason (2003, p. 115) reminds us, the 'idea that market forces, if left to themselves, can solve economic problems is not new – it lost credibility in the 1930s'. In the long term, the viability of particular primary and secondary industries will be jeopardised and lifestyle options will contract as available product ranges diminish.

Economists and other specialists whose knowledge and opinions underpin political decision-making consider market logic a rational outcome of supply and demand choices and innovation. This creates a tendency to view 'the market' as an entity characterised by product consumption or demand led pricing (Anadon and Holdren, 2009; Bazerman, 2009; Pasternak, 2000). From the perspective of energy consumption, or pricing that takes fuller account of the real costs of production, including their greenhouse gas emissions, such practices lack logic or display unusual expressions of it. These features of energy and climate change debates impact directly upon the prospects of prosperity for many people and the abilities of many to participate in the international political economy.

Political and economic imperatives differ and produce diverse social structures and views of the future. Both politics and economics seek to comprehensively order human activities and normalise values, aspirations and conceptions of what is desirable. Economic and political processes and relationships reinforce and privilege certain forms of production over others, and in the 20th century increased the scale of global greenhouse emissions. Recognising politics, rather than economics, as the principal driver of climate change responses offers greater opportunities for equitable, widely supported and ethically just solutions.

Water, population and security

Water resources are important components of the natural environment and also of the built environments, residences, industries and centres of production associated with human societies. Modern industrialised societies have been made possible by their use and exploitation of water resources (Baron et al., 2002). The formation of major sites of production, and dense centres of population to provide labour forces and ready markets for manufactured commodities, relied upon accessible, affordable and large volumes of water. Modern industries are heavy users of water and without readily affordable large water volumes, many industries are likely to lose their economic viability. Even in terms of domestic household use, per capita water consumption is often higher among those living in the urban centres of industrialised states as a result of heating, lighting and other lifestyle factors. As reported in 2005 by the United Nations Educational, Scientific and Cultural Organization:

> [w]orldwide, 70% of the water that is withdrawn for human use is used for agriculture, 22% for industry and 8% is used for domestic services. In general, these proportions vary according to a country's income: in low- and middle-income countries, 82% is used for agriculture, 10% for industry and 8% for domestic services. In high-income countries, the proportions are 30%, 59% and 11%, respectively... It is estimated that the average person in developed countries uses 500–800 litres of water per day (300 m^3 per year), compared to 60–150 litres per day (20 m^3 per year) in developing countries.
> (United Nations Educational, Scientific and Cultural Organization, 2005, http://www.unesco.org/water/news/newsletter/92.shtml#know)

88 *Climate Change and Order*

Without secure and predictable rainfall and the infrastructure necessary to store and distribute water, the economies of developed states may cease to function in currently recognisable or orderly forms.

Growing urban populations also place increased stresses on water availability (Strzepek and Boehlert, 2010). Throughout the 20th century, population growth and rising living standards shifted diets towards more water intensive foods. The spread of industrialisation and urbanisation increased water usage for energy generation. In the 21st century, these factors, coupled with the potential effects of climate change on the hydrological cycle are increasing the prevalence of government policies that seek to reconcile the needs of ecosystems, people and industries. Reconciling the competing claims of urban populations and healthy river systems with the need for agriculture is not an easy matter. The net 'effect of these demands can be dramatic in key hotspots, which include northern Africa, India, China, parts of Europe, the western US and eastern Australia among others' (Strzepek and Boehlert, 2010, p. 2927). As has been observed, while '[c]arbon is a measure of the anthropogenic causes of climate change; water is a measure of its impacts' (United Nations Educational, Scientific and Cultural Organization, 2009, p. i).

In the near future, water changes will further magnify gaps between the capacities of developed and developing states as they seek to respond to global climate change (Roberts and Parks, 2007). As has already been noted, at least in the short term, the most severe impacts of water redistributions will be experienced by developing states, largely because their financial, technical and material capacities for mitigation and adaptation are more limited. However, as the costs of implementing appropriate policies, mitigation and adaptation strategies impact upon societies, many developed states will find themselves considerably weakened by expanding demands upon their administrative and financial resources.

Water is utilised in producing electricity, and almost every imaginable form of industrial production. Its availability depends upon government policies, infrastructure, the actions of economic sectors and a complex series of natural events. The volume and quality of available water resources in any given location is subject to global climatic patterns; government policies and economic activities; and an array of actions by others (Weltzin et al., 2003). Even aside from increasing water challenges, developing states presently lack wealth, investment-led economic growth and widespread industrialised production. In spite of these difficulties, many hold aspirations of economic growth and nurture visions of increasing affluence that include secure freshwater supplies for households and longer life expectancies for citizens.

Water redistribution challenges the authoritative capacities, industries and economic sectors of all the world's states and will occur in many different ways in different locations (Vörösmarty et al., 2000). They will also fluctuate quite considerably over seasonal cycles and be unevenly distributed within the territories claimed by individual states. In the 21st century, many states with highly urbanised populations will confront enormous political and economic challenges in seeking to secure adequate water resources to support their city-dwelling peoples and industrial production, agricultural activities and energy needs (Vörösmarty et al., 2000). Even with sustained high levels of political will and well-targeted national and international water policies, these problems will pose significant challenges.

An important political ramification of these complexities is that they create poor negotiating contexts for decision-makers (Hollins, 2010; Edmondson, 2001). Among other things, these conditions tend to produce negotiating dynamics whereby short-term and single-issue coalitions form among states but lack the flexibility and durability to become lasting agreements (Stokke and Vidas, 1996; Young, 1994). They create uncertainty, reduce the durability of international agreements and jeopardise their effective implementation. These dynamics were evident in delaying the implementation of the Kyoto Protocol and in limiting political agreement at both the Copenhagen Climate Change Summit in 2009 and the Cancun Conference of Parties in 2010.

Water redistributions are likely to sever the ideational and ideological links between historically accepted paths of development and progress, increasing economic growth, further industrialisation and affluent lifestyles. At any point in history, changes to the availability, quality and distribution of water resources arising from global climate change would have held significant implications for the international political community and many of the peoples and states within it. That these changes are now occurring in a context of highly interdependent global markets, globalised production practices and increasing populations, necessarily intensifies the scale of anticipated political and economic disruptions (Vörösmarty et al., 2000). These contextual complexities increase the likelihood of international conflict arising from water redistributions (Homer-Dixon, 1994).

Politically, the consequences of global climate change highlight the importance of international conflict management strategies, since water shortages are likely to become a key motivation for resource conflict

(Homer-Dixon, 1994; Gleick, 1993). With more than 250 transboundary river basins, conflicts over water may be one of the principal threats to international security in the 21st century (Vörösmarty and Sahagian, 2000, p. 763). Water shortages, floods and droughts, changed river flow patterns, erosion, dust storms and fires all pose both direct and indirect threats to many human societies. Where and how people live, what they eat and wear and how they protect themselves from threats, the illnesses they suffer and their prospects of recovery are all directly impacted by the availability, quality and security of water resources (Vörösmarty et al., 2000).

Many of the world's poorest people will be dramatically impacted by water redistributions. Efforts to reduce global poverty and child mortality are likely to be negatively impacted by the increased incidences of floods, land inundation and water contamination. Freshwater shortages dislocate populations, reduce regional agricultural production and undermine fledgling industrialised enterprises (Bates et al., 2008). In summary, '(f)uture climate change poses one of the greatest threats to poverty eradication' threatening the survival prospects of many people, and reducing the likelihood that dreams of global prosperity will be fulfilled (Dow and Downing, 2006, p. 57). Water-related challenges arising from climate change highlight the extent to which '[h]uman security is a multi-faceted concept', wherein security, survival, human rights and responsible governments comprise linked but incomplete variables (O'Neill, 2009, p. 43). Even the most effective and responsible of governments cannot make the rain fall, and there are limits to their abilities to re-route rivers. As vulnerabilities increase, armed conflict might also increase as people experience '[d]rought, water scarcity' and 'land degradation' (O'Neill, 2009, p. 44). In the period 1820–2007, over 400 international agreements were achieved between states regarding 'international freshwater resources, where the concern is water as a scarce or consumable resource, a quantity to be managed, or an ecosystem to be improved or maintained' (Transboundary Freshwater Dispute Database, http://www.transboundarywaters.orst.edu/index.html). This number is likely to increase in coming decades as a consequence of efforts to avoid disorder and disrupted development.

Industrial and urban dependencies

Industrialisation made it possible to produce unprecedented amounts of food, including in areas that would not have been agriculturally productive without artificial irrigation, capital-intensive forms of production, scientifically enhanced crops and manufactured fertilisers and pest control (International Food Policy Research Institute, 2002). Industrialisation also made it possible to form and maintain large-scale urbanised human communities. These benefits of industrialisation will be increasingly difficult to sustain as global climate change alters water distributions. The plausible consequences of climate change–related water redistributions include the disorderly redistribution of human populations (Vörösmarty et al., 2000).

Modern, developed, industrial societies are underpinned by practices that have improved the lives of the majority of their members and often sustained the lives of their most vulnerable. These achievements are notable, commendable and highly desirable but are increasingly perceived to have arisen from unsustainable manipulations of the natural environment (Wackernagel et al., 2002). Against this backdrop states have begun to search for and develop policies that offer globally equitable mechanisms to secure future social, political and economic development. These initiatives are motivated by the weight of accepted science, and an increasingly unified and strident scientific community, who believe:

> The planet is warming due to increased concentrations of heat-trapping gases in our atmosphere...Most of the increase in the concentration of these gases over the last century is due to human activities, especially the burning of fossil fuels and deforestation...Natural causes always play a role in changing Earth's climate, but are now being overwhelmed by human-induced changes...The combination of these complex climate changes threatens coastal communities and cities, our food and water supplies, marine and freshwater ecosystems, forests, high mountain environments, and far more.
> (Gleick et al., 2010, p. 689)

Addressing climate change, the causes and effects, then require fundamental changes to currently accepted lifestyle practices and aspirations.

The international political dimensions of changed water and rainfall distributions are complicated by regional variations and the relatively low levels of detailed, accurate and readily accessible knowledge concerning these climate change consequences (Dow and Downing, 2006; Vörösmarty et al., 2000). While a general level of knowledge exists concerning changing ocean water volumes and freshwater supplies, these often fall short of enabling meaningful projections to be established, and are inadequate bases of strategic planning for adaptation or mitigation (Giordano and Wolf, 2003). Although some doubt remains concerning specific details of current risk assessments from rising global temperatures, these are much more clearly understood than the water-related impacts of global climate change. The complex interplay between ocean currents, tidal surges, storm formations and rainfall patterns across the world complicate future projections. In a sense, this is not surprising because many scientists have paid close attention to atmospheric changes over several decades to monitor and understand the effects of elevated levels of greenhouse gases. The same level of sustained scientific attention has not been directed by the scientific community towards water issues and political leaders historically conceived of these as territorially based resources.

The intensive energy and water consumption of cities is readily apparent and easily explained. Constructing, maintaining and sustaining the infrastructure of cities to ensure they function as centres of enterprise, commerce and culture consumes vast quantities of electricity and other forms of energy (United Nations Human Settlements Programme, 2008). Their high needs for water consumption, however, are both more challenging and less visible (Roaf et al., 2005). For instance, within industrialised states, there is a tendency for drought to be regarded very largely as a weather phenomenon that impacts upon farmers and/or rural communities rather than being perceived as a potential threat to an entire population. When cities confront potential water shortages, people look to their governments for solutions to alleviate prospects of impending hardship. These contemporary political realities reflect a significant historical change because in the past, when water relocated, the people moved. Now, developed states expect to keep their cities in situ and to shift water (in and out) to meet their needs. Doing so necessitates extraordinary levels of infrastructure investment as nature is subordinated to the demands of human communities.

Water resources form part of the intrinsic environment and social/economic backdrop to the political contexts of human societies. Providing secure and privileged access to water has been central to the establishment and maintenance of state borders, and other claims to authoritative status over territories, peoples and resources (Blatter and Ingram, 2000). The water that flows through territory has been subject to state-based rights of regulation and resource allocation that enabled agricultural irrigation, production of hydroelectricity and other contributions to economic and social well-being (Sax, 1990). States, and the private economic actors they host, have presumed that water will continue to be available for their use and have premised their long-term interests upon its continuing availability. Water redistributions arising from climate change, which jeopardise the availability of water for future needs, can then be understood as a fundamental challenge to the viability of states.

International law provides a means of dispute settlement when states need to reconcile competing claims to resources, including regulating their access or use of water resources (Transboundary Freshwater Dispute Database, http://www.transboundarywaters.orst.edu/index.html). International law also provides a means of addressing ongoing environmental damage when the actions of one state jeopardise the resource access abilities of others, such as in transboundary air or water pollution. The possibility that climate change might cause rivers to dry or re-route has not yet been taken into account in such negotiations or agreements between states. This new 21st-century reality holds particular political importance because the absence of adequate water resources directly threatens the well-being of states and is likely to provide a key source of conflict, notwithstanding the likely emergence of new international water markets (United Nations Educational, Scientific and Cultural Organization, 2009; Blatter and Ingram, 2001).

Adaptive capacities

Developed and developing states have different capacities for dealing with their particular climate change challenges and these are reflected in the mitigation and adaptation strategies they attempt to develop, or support within the international political community. In a very real sense, it is their individual experiences of addressing the challenges of floods, coastal and river inundations, droughts and desertification that drive their participation in international climate change discussions and negotiations. The United Kingdom, for instance, believes that its impending shortages of fresh water might be addressed by importing it from elsewhere in Europe (Cohen, 2008). While this solution might never be implemented, or might prove less viable than anticipated, the fact that it has been the subject of serious consideration reflects the history of state-based problem-solving, progress, financial and infrastructural capacities. It illustrates ongoing beliefs that the challenges confronted by individual states can be addressed by drawing upon resources from other states and that neighbouring states will make their resources available. By contrast, even if the states of sub-Saharan Africa had near neighbours with water to spare, they lack the financial capacity to purchase it and the infrastructure necessary for transporting and distributing it. In climate change management, mitigation and adaptation, financial, infrastructural and social adaptability are key determinants of possible opportunities and the unequal distribution of these capacities.

The international political community confronts a further layer of political challenges arising from the impacts of changed rainfall, rivers and rising sea levels upon developing states. These are likely to disrupt the promises of development that many developing states have aspired to achieve in recent decades because industrialised production processes utilise large volumes of water and many developing states are likely to be adversely affected by changing global rainfall patterns (United Nations Development Programme, 2008; Roberts and Parks, 2007). For states that are currently struggling to address high rates of poverty and low rates of industrial development, the impacts of water-related climate change are likely to end their prospects of prosperity and progress. These issues present major political and security problems for the international political community. Even if the ethical matters associated with persistent and future inequities could be sidestepped, the fact developing states are highly populous makes it impossible to ignore the problems they confront.

Conclusion

'We basically have three choices – mitigation, adaptation and suffering. We're going to do some of each. The question is what the mix is going to be. The more mitigation we do, the less adaptation will be required, and the less suffering there will be' (John Holdren, President of the American Association for the Advancement of Science cited by Kanter and Revkin, 2007).

Global climate change is altering almost every imaginable facet of the physical contemporary world, including aspects that are directly and indirectly crucial to current and future human societies. Growing awareness of the breadth of these impacts makes addressing climate change complex and intimidating. The diverse effects of global climate change are altering the range of temperatures, the nature of yearly seasons, the range of land and water dwelling species, the flow of rivers and the internal currents of heat and cold exchange that drive the oceans (Nicholls et al., 2007b; Flannery, 2005). Climate change is altering the range of locations that will be suited to human habitation into the future and the possible locations in which agricultural activities might be fruitful (International Food Policy Research Institute, 2009; United Nations Development Programme, 2008). These observations would present enormous challenges to any configuration of societies, political systems and forms of economic enterprise. The reliance of many contemporary political systems and their economies upon wide ranges of water intensive enterprises significantly increases the potential upheaval they will experience as rainfall patterns change, water resources are redistributed and sea levels rise.

In the past, states applied common heritage of humankind principles in their efforts to conserve, protect and regulate the use of ocean and seabed resources. More than other aspects of the physical world the oceans have been regarded as 'there for everyone' – including land locked states. In numerous international discussions and agreements, a key objective has been equitable access to the oceans, seabeds, their resources and opportunities (Giordano and Wolf, 2003). Many of these agreements have integrated environmental, economic, social and security issues, reflecting the breadth of their political importance.

Following repeated failures to establish internationally endorsed climate change mitigation and adaptation strategies, the international political community might draw upon these lessons. The extent to which historical relationships continue to be important influences upon political dynamics and processes for negotiating and implementing international agreements should be borne in mind as states prepare for new rounds of climate change and water resource discussions. Revisiting these dynamics might help to generate alternatives to the logjams and decision-making impasses that otherwise seem inevitable. Among states, there is a very real likelihood that new policies will continue to reflect earlier historical patterns of environmental neglect on the one hand, and resilient resourcefulness on the other (O'Brien and Leichenko, 2000).

It will take democratic states time to adjust the relationships among political, economic and social actors, and the networks of rights and responsibilities that occur within them (Adger et al., 2009; Achterberg, 2001). Ultimately, these relationships will be transformed by the combined threats arising from the security implications of global climate change and the impacts of rising temperatures, sea levels and more intense weather events. The political histories of human societies reveal their sensitivities and vulnerabilities to security threats, and demonstrate their potency as sources of changed attitudes and practices. The current emphasis in favour of lengthy negotiations and multilateral climate change agreements might then be abandoned in coming decades. If the lessons of the past provide meaningful indications of the likely political impacts of climate change, we might anticipate sudden disruptions to present patterns as climate-related security risks emerge.

6
Energy, Progress and Population

Introduction

At this point in the 21st century, states are confronting changes in their physical environments and the conditions upon which their economies and systems of political and social organisation are based, and many are clinging to historically established aspirations, expectations and modes of behaviour. For these states, the changes that are occurring in the fertility and habitability of their land, their access to secure freshwater resources and energy supplies are unfolding alongside persistent hopes of prosperity. Their aspirations are sustained by enduring beliefs that modernisation and industrialisation are synonymous with progress, and corresponding fears that relinquishing such aspirations might threaten the well-being of existing societies and structures. For many, their abilities to enjoy long lives, access to secure food supplies and advanced medicines are linked with the overall economic prosperity attained through globalised production and economic markets in the 20th century. These achievements occurred through successive international arrangements and political structures that were premised upon energy consumption.

Those who engage in efforts to address the causes and consequences of global climate change confront political complexities arising from the composition of the international political community. The reality that there are many more developing states than there are affluent and industrially developed states presents an array of political problems that are exacerbated by population and global energy demands. Many developing states are attempting to attain unprecedented levels of economic prosperity for their rapidly expanding populations, seeking industrial development and new roles in the international political community.

Those developing states that are not striving to create affluence are, at least, seeking to reduce the high incidence of poverty that is experienced by most of their people, and all states are attempting to reduce their economic and political vulnerabilities through their involvement in the international political economy (O'Neill, 2009; Paterson, 2000).

Observing that patterns of production and consumption in fossil fuel energy sources are at the heart of human induced climatic changes can promptly lead to dead ends in policy developments and political negotiations (Stern, 2009; Parry et al., 2007). The connections between industrial development, prosperity and environmental damage, including climate change, make meaningful mitigation strategies tremendous political challenges (Anderson and Leal, 1998). Minimising the impacts of global climate change would be difficult under any circumstances, but these political challenges are exacerbated by the inequalities of circumstance and the political costs of foregoing short-term economic aspirations (Roberts and Parks, 2007; Yohe et al., 2004). The developed democracies tend to enjoy greater levels of political stability than developing states and developed states are geographically removed from most of the flows of refugees seeking to escape from violence or disaster. These trends contribute to the complexity of equitable and ethically just responses to climate change (Gardiner and Hartzell-Nichols, 2012; O'Neill, 2009; Roberts and Parks, 2007).

Established energy intensive forms of industrialisation within developed states have made the lifestyles they support the envy of many people. Their consumption of energy resources enables those who live in urban centres to earn more than those engaged in rurally based employment, enjoy better health, and easier lifestyles (United Nations Environment Programme, 2005). The fact that more people than ever before wish to emulate these developmental paths, in order to live these lifestyles, presents the global community with a significant set of problems (Roberts and Parks, 2007). The earth's resources and ecosystem, the planet's natural capital, cannot support or meet these lifestyle aspirations for a growing population, presenting political, economic, moral and governance problems for the contemporary world (Gardiner and Hartzell-Nichols, 2012; United Nations Development Programme, 2008; Dryzek and Schlossberg, 2005). The planet's natural capital is insufficient to support the elevated resource demands of an expanding global population, and this presents unprecedented political challenges.

Contemporary industrialised societies have established themselves through their intensive consumption of an array of fossil fuels. These are utilised in constructing buildings, infrastructure, factories and the industries that support their production (Roaf et al., 2005). They are used in producing and packaging products for sale and in delivering them to their respective marketplaces. They are utilised in the wholesale and retail enterprises from which they are purchased, and in the homes, businesses and other locations in which they are finally consumed. Fossil fuels are used in heating and lighting the homes, hotels, hospitals and other residential facilities in which the populations of developed states reside, and in the tall towers of commerce, the regulation of urban traffic flows and the conduct of governmental affairs. They are used in producing the food and clothing that sustains contemporary communities, and in their recreational pursuits. These realities pose challenges for developing policies to mitigate the worst impacts of global climate change and/or to support human, economic and social adaptation to its consequences (Prugh et al., 2008; Morse and Myers, 2001).

In developing states, patterns of fossil fuel consumption sometimes reflect firmly entrenched class divisions or other forms of economic and social privilege, including access to political participation. In these instances, wealthy minorities enjoy the ease of their affluence in motor vehicles, electricity and access to globally marketed products although the majority of their compatriots lack access to such commodities. For some developing states, most notably in the Middle East, their ability to participate in the international political economy relies upon the oil, gas and other energy resources they can export to an energy hungry industrialised world. These states face political and security challenges arising from widespread poverty and many are also directly vulnerable to climate change impacts such as increasing frequency of droughts (Page, 2006; Adger et al., 2003). Arguably, the political challenges of reducing reliance upon fossil fuel energy sources and/or reducing the greenhouse gas emissions associated with their use are even greater for developed states (Umbach, 2010).

Political solutions and effective leadership will be necessary to manage the issues presented by global climate change and energy demands in the 21st century. Negotiating political solutions to changing international patterns of energy production and consumption will require sustained recognition of the potentially disparate and uneven economic and social impacts that will arise from meeting internationally agreed greenhouse gas emissions targets. In the globalised economy of the 21st century, the reality that the overall well-being of many people in the contemporary world derives both directly and indirectly from access to affordable energy supplies is politically complex. Part of this complexity arises from the competitive pursuit of economic growth, as states 'compete for market shares for their firms and regions in the world market' while 'promot[ing] and attract[ing]...high profit firms and industries' (Inayatullah and Blaney, 2004, p. 129).These interrelated political goals inevitably create competition between states even when they share a common vision or common risk assessment in relation to their abilities to protect their people and territories (Manne and Richels, 1995).

Politics and energy consumption

Developed states are not the sole utilisers of fossil fuels but their rates of consumption greatly exceed those of developing states (Flannery, 2005). It is an important factor in climate change discussions and negotiations that developed states consume approximately 80 per cent of the fossil fuels that are utilised worldwide annually (Pinderhughes, 2004). While the fuel utilised in developing states also contributes to global emissions, the energy forms they most commonly utilise often produce lower rates of greenhouse gases as the fuels for cooking and heating derive largely from raw energy sources such as wood and manure rather than fossil fuels (Roberts and Parks, 2007). Also, lower levels of energy consumption in developing states lead to lower overall levels of carbon emissions. These disparities in fossil fuel consumption between developed and developing states form an important site of international debate and ongoing political discussions (O'Neill, 2009; Page, 2006).

In recent years, global greenhouse gas emissions have been the principal site of widespread political exchange and negotiation concerning resource- and energy-related issues. As government leaders prepared for the Copenhagen negotiations that occurred late in 2009, they attempted to reconcile immediate economic and electoral consequences while seeking to reduce prospects of longer term potential upheaval. They engaged themselves in complex political dynamics with others, but failed to achieve a clear vision of the future. After more than a decade of intensive international political debate, emissions targets remain largely unresolved because of continued aspirations for economic and industrial growth and security (O'Brien and Leichenko, 2000). Disparate patterns of industrialisation, affluence and stalled economic development among states are important contextual factors because these differences influence the forms of greenhouse gas emissions reductions targets that states favour and the monitoring and compliance mechanisms they are prepared to accept or impose upon others (Roberts and Parks, 2007).

Tackling shared vulnerabilities

Climate change debates expose the vulnerabilities of political leaders who are reluctant to establish firm and meaningful greenhouse emissions targets supported by policies to change patterns of energy production and consumption (Brauch et al., 2008; Barnett, 2001; Mathews, 1991). International debates provide a new operating context for the legislative and regulatory efforts of contemporary governments, and in the competitive context of energy debates there are 'at least as many policy instruments as...market problems demanding the attention of policy makers' (Tinbergen, 1956 in de Coninck et al., 2007, p. 338). In international climate change discussions, this broad 'menu of policies' increases political complexities (de Coninck et al., 2007, p. 351).

In the 21st century, debates concerning energy production and consumption are confused by broader debates concerning the economic and political implications associated with climate change mitigation and adaptation issues (Pasternak, 2000). This occurs because access to energy resources and the ability to ensure prosperity and security are inextricably and interdependently linked. The goals of prosperity and security that lie at the heart of contemporary governments and the expectations of their citizens give particular intensity to these dynamics in debates concerning energy. They thereby limit the abilities of governments to pursue energy strategies that are markedly different from the energy forms and market structures that are already in place (Bazerman, 2009). Internationally, these dynamics add to the 'too-hard' nature of achieving effective emissions reductions and other climate change mitigation strategies.

The international political community faces real and ongoing political challenges in seeking to alleviate some of the likely political consequences of global climate change impacts (Borgerson, 2008). The slow rate of progress towards meaningful targets, international emissions agreements and changed energy consumption is of increasing concern as predictions of inevitable long-term climate change impacts become more serious. Every year of delay commits us to increasingly elevated global temperatures that increase the risks of less predictable and more unmanageable climate change threats to human societies (Diffenbaugh and Scherer, 2011; Battisti and Naylor, 2009; Tubiello et al., 2007; Weltzin et al., 2003). Notwithstanding the seriousness of these risks, the political dynamics at stake make it unhelpful to focus criticisms upon particular states.

The forms of government that have coincided with industrialisation and 20th-century prosperity have grown accustomed to unilaterally responding to political, economic and social problems within their borders by expanding their possible policy options, and these expectations are unhelpful for international climate change management. While governments endeavour to establish common goals in relation to global climate change management, they are ham-strung by their histories of competition and the competitive markets they have supported (Eakin et al., 2009). Finding new ways of reconciling their localised interests with universal imperatives in climate change mitigation and adaptation strategies demands new ways of responding to established political dynamics (Edmondson, 2011). To date, the universalising tendencies of liberal democracies have proven unhelpful in this regard. This is partly because their commitments to electoral popularity alongside capital-driven economies renders them poorly equipped to demonstrate international leadership by taking domestically costly decisions to reduce emissions and fossil fuel energy reliance.

Politics before policies

Presently, 'governments face the difficult task of balancing the upfront costs of mitigation and adaptation against the risks of climate change itself as well as the risks of having to rapidly reduce emissions in a way that could prove highly costly' (Gallagher, 2009, p. 20). Collectively and individually, states must come to terms with the realisation that they only have limited choices to ensure the longevity of their societies (Gallagher, 2009, p. 20). They must also recognise that their short-term policy choices also determine the range of choices that continue to be available into the future (Eakin et al., 2009; Eckersley, 2004).

In responding to the political challenges of global climate change, states must confront the need for global mitigation strategies that unfold in successive stages. These will require them to take actions to reduce 'the pace and magnitude of the climatic changes being caused by GHG emissions' (Gallagher, 2009, pp. 20–21). Across the international political community, states need to alter their patterns of energy use and establish new cornerstones of international and domestic adaptation strategies. In the 21st century, effective international responses will commit states to reducing their greenhouse gas emissions in order to avoid the most adverse impacts of climate change and to reduce their risks of energy-related conflicts (Evans, 2010). Governments and the international organisations that support their climate change efforts will also need to develop policies to manage the 'suffering' that will be experienced from the climate change impacts that are 'not averted by either mitigation or adaptation' (Gallagher, 2009, pp. 20–21).

As already argued, to date, states hold rather poor records in this regard. Their limited policy capabilities are exacerbated by their simultaneous needs to direct political and economic resources to mitigation and adaptation strategies. However, the more emissions that can be reduced in the near term, the more likely that deep and widespread suffering can be avoided (United Nations Development Programme, 2008; Roberts and Parks, 2007; Flannery, 2005). Even if effective mitigation strategies can be successfully agreed upon and implemented in a timely manner, societies, political and economic systems will also 'need to adapt to the inevitable climatic changes' that will occur into the future (Gallagher, 2009, pp. 20–21). The international political community is, therefore, challenged by the need to respond to global climate change by intervening in previously autonomous decisions concerning the use and distribution of energy sources and resources, the forms of energy production and the technologies that underpin them (O'Neill, 2009; Rudolph, 2005).

The historical relationships between people, their societies and energy resources form an important context for the international political community as responses to global climate change become increasingly urgent (Eckersley, 2005; Paterson, 2000). They contribute to a tendency for governments to drag their heels or to be ad hoc in pursuing national climate change mitigation strategies, taking one step forward, two steps back, instead of exercising political leadership (Edmondson, 2011; Stern, 2009). While 'wait and see' political dynamics are also apparent in other international political dealings among states and considerations of global climate change issues, an absence of international political leadership now seems most pronounced in relation to energy issues. There is a very real need for political leadership within the international community to establish new patterns of energy use, supported by new policy developments and implementation practices. These will challenge and potentially overturn approaches to policy-making that rely upon popular support and the cooperation of major economic actors.

Appropriate and effective international agreements and regulatory mechanisms necessitate international agencies with the authority to ensure that states and other actors comply with international agreements and established targets, whether these are domestically or internationally determined (Kütting, 2000). These will need to be equipped to respond to the tensions and political complexities that arise between international- and state-based policy and regulatory contexts. Without established and recognised authority structures, international agreements and enabling agencies will struggle to support mitigation and adaptation efforts. Without them, states will seek to avoid political dilemmas and produce agreements that fail to be universally implemented, or lack durability (see, for instance, discussions concerning the Kyoto Protocol in Chapters 2 and 12).

Effective decisions must recognise that throughout history, states have held individual, nationally located rights to exploit the energy sources available within their territories and many continue to see this as a basic political right. States have not required permission from others in order to utilise energy sources as they have seen fit. Global climate change challenges this historical prerogative and it is highly likely that climate change responses, coordinated at an international political level, will seek to mediate or prohibit this customary right of states (Conca, 2005). Effective international responses to climate change will rely upon a 'global deal' that is 'effective', 'efficient' and 'equitable' (Stern, 2009, p. 4). Such a deal will reduce 'emissions on the scale required;... keeping costs down; and... taking... account [of] both the origins and impact of climate change' (Stern, 2009, p. 4).

In some contexts, international political embarrassment or isolation can be expected to produce changed behaviour or actions by a particular state. However, climate change–energy issues are such that risks of embarrassment are unlikely to persuade states to abandon their goals of economic progress and prosperity for the intangible rewards of good international citizenship. As has always been the case, states will act when they perceive an urgent and direct threat to their well-being and they can identify possible solutions that bear familiar structural hallmarks (O'Neill, 2009; Young, 1994). Many states will also want to know what others are doing before committing themselves to any possible actions. Some states fail to engage effectively in international policy negotiations concerning the distribution of relative costs and responsibilities in responding to global climate change imperatives because they cannot resolve their internal dilemmas or identify solutions they can be confident of implementing (Young, 2002; Luterbacher and Sprinz, 2001). For instance, it is relatively straightforward, at one level, to argue that the United States ought to exercise stronger political leadership on the grounds that it 'has the largest economy, uses the most energy (and within that total the most oil), has made the largest cumulative contribution to the atmospheric buildup of fossil carbon dioxide... has a large balance of payments...and stake in...security benefits of meeting global energy needs' (Anadon and Holdren, 2009, pp. 89–90). Nonetheless, such an argument misses the central point concerning the importance of sovereign authority and the imperatives towards a continuing independent quest to consolidate economic security through ongoing industrial progress.

Similarly, we might argue that as the country with the highest per capita greenhouse gas emissions, Australia has a particular responsibility to set clear targets and/or impose a carbon tax on emitters (Flannery, 2005). However, such choices are highly problematic for Australia's federal government for reasons that go beyond straightforward electoral unpopularity. Australia has a relatively small but geographically dispersed population which means that there are relatively high energy-consumption rates incurred by transporting resources to production and processing sites, and distributing goods across this dispersed population. Additionally, Australia's economy is currently heavily dependent upon large volume exports of key mineral resources, including coal, which increases the difficulties of taking strong decisions to reduce domestic energy consumption, without also choosing to reduce coal and other energy resource exports. The electoral and economic costs of simultaneously reducing both domestic energy consumption and exports would be overwhelming.

New energy policy drivers

Energy-related resource use is so fundamental to modern industrial societies that it makes the politics of international climate and energy policies especially intense. International energy and climate change management policies are often poorly constructed and directed towards confused social and economic goals. In the 21st century, responding to the political challenges arising from energy use and distribution, and climate change mitigation requires predictable and durable forms of cooperation from diverse governments to reduce these confounding political dynamics (Edmondson, 2011; Vogler, 2005; Young, 2002;). Including diverse economic and political interests will support international negotiations and expand shared knowledge of new forms of energy production and technologies because these actors are interested in preserving their long-term markets (Stern, 2009; Sawin, 2003; Kütting, 2000).

Private economic actors are important drivers and contributors to the domestic policies pursued by governments (Alfsen et al., 2010;

Anadon and Holdren, 2009; Van Noorden, 2008). As global climate change unfolds, they will also shape the policy contexts of international energy agreements. Optimistically, it might be hoped that their future roles will include 'financing the additional investment required' to enable the development of viable alternative commercially available energy sources (United Nations Committee for Development Policy, 2009, p. v).[1] Nonetheless, the framework decisions taken by states and the agents who act on their behalf will retain primary political responsibilities for the energy production, consumption and distribution patterns that emerge. Avoiding compounding existing inequalities between developed and developing states will only be possible if governments, intergovernmental organisations and economic actors establish climate change mitigation strategies. These will be most effective if they encompass the goals and influence of diverse parties, including public sectors, and proceed through international cooperation (United Nations Committee for Development Policy, 2009).

Energy economics

It is in the realm of energy supplies and access to appropriate energy resources that climate change debates are most intense – and yet least able to generate effective international agreements. It is also in this context that the complex interplay between economic and political interests and imperatives among states are most marked. These tend to occur in ways that are unhelpful to clear and effective decision-making. It is in this area of climate change debates that the costs of pursuing or adopting effective mitigation and adaptation strategies become most starkly apparent. It is in relation to energy resources that the costs of global climate change become most evident to most people, resulting in mixed and sometimes conflicted interests.

Lurking within the political problems arising from climate change are challenges concerning the nature of political authority as governments face new uncertainties in seeking to preserve the well-being of their people (Morse and Myers, 2001). Changing aspirations and expectations regarding possible lifestyle options is part of this complex of political issues. These are exacerbated by the need for new expectations

---

[1] Such hopes are buoyed by the roles of private economic actors in protecting vulnerable European states, including Greece, Spain and Ireland, from economic collapse following the Global Financial Crisis 2007/2009 and its continuing repercussions through 2011/2012.

and aspirations concerning effective government, the representation of their interests, and ways of accommodating diversity. In recent decades, climate change debates have exposed fault lines in these political parameters, especially among developing states who argue in support of their resource conservation even as they claim political rights to pursue economic progress (Eckersley, 2006; Page, 2006).

In relation to long-term climate change–energy issues, these fault lines highlight the political sensibilities of states, and the challenges they face in reconciling competing interests (Evans, 2010). States are responsible for ensuring the protection and defence of their people and territories and this relies upon maintaining borders, defence forces and other sources of law enforcement. Each of these functions of sovereign authority relies upon the availability and allocations of material resources, including energy supplies. In this, states inevitably play 'contradictory role(s)...facilitating both...destruction and...protection' of the resources that underpin their prospects of economic growth and territorial security (Eckersley, 2005, p. 159).

The energy elephant

One of the elephants in the room of international energy debates is the basic reality that meeting the expanding energy needs of a growing world population, and supporting new levels of production and economic prosperity among developing states, will not be possible into the near future (United Nations Committee for Development Policy, 2009; Mason, 2003). By 2002, increasing numbers of scientists were arguing that human demands on the environment 'may well have exceeded the biosphere's regenerative capacity since the 1980s' (Wackernagel et al., 2002, p. 9266). In effect, they were arguing that at least two decades of over-exploitation of natural resources and habitat destruction had created an inevitable tipping point towards lasting environmental damage. Additional energy burdens are forecast since it is anticipated that by 2020, approximately '60 per cent of the world's population will be urban' with dramatically increased energy demands (United Nations Environment Programme, 2005). Some agencies, including the International Energy Agency, predict that between 2020 and 2030 global demands for energy will increase by 60 per cent (United Nations Educational, Scientific and Cultural Organization, 2009). These basic political realities, which might be summarised as 'over-population and over-consumption', make it more difficult and more urgent for effective international climate change mitigation and adaptation strategies to be developed and rapidly implemented (O'Neill, 2009, p. 42).

It is also worth bearing in mind that the 'mechanization of agriculture and dependence on petrochemical fertilisers and pesticides largely created the green revolution', enabling a smaller number of people to produce sufficient food for an expanding population in the 20th century. There is a level of truth in claims that 'modern agriculture uses land to convert petroleum into food' and any decline in petroleum and its by-products will compromise the capacities of global agricultural production (Mason, 2003, p. 9). Reduced fossil fuel energy consumption in agriculture will, at least in the short term, substantially reduce global food production (Mason, 2003). Some scientists predict that when oil supplies dwindle 'world agriculture will be able to provide food for many less people – on one estimate, only 2 billion' and this will impact upon a world that is expected to have 8 billion people by 2030 (Mason, 2003, p. 9).

Currently, possible solutions to these problems entail either greater inequities in food distribution, food poverty for a greater percentage of the global population, or a fundamental reappraisal of what constitutes food security. In each case, the costs of food will rise significantly and some economists argue that we have already passed this tipping point (Friel et al., 2008; Jolly and Ray, 2007). As these realities bite more fully into international negotiations, and production and consumption practices change, they will further sharpen and problematise distributions between developed and developing states. Their impending disruption to collaborative and cooperative endeavours to combat climate change increases the urgency of prompt political actions.

Political difficulties

In international climate change and energy debates, the individual and collective patterns of production and consumption of energy resources become direct drivers of states' policy interests. This means that as states enter into discussions concerning greenhouse gas emissions, they are also interested in maintaining their economically viable and prosperous industries and securing the prosperity of their people. They are interested in representing the interests of their citizens and economic actors within the international political community, while also seeking to influence the decisions and policies of other states and foreign economic actors (Edmondson and Levy, 2008). For many states, these issues are closely linked with electoral sensibilities concerning prospects of continuing government stability and prospects of re-election.

Developed states face considerable political and policy challenges since all of their production and consumption practices rely upon a variety of direct and second-generation forms of energy consumption. Their cities, housing, infrastructure and industries (including food production) are premised upon their abilities to produce and consume energy (Gallagher, 2009; Roaf et al., 2005). These circumstances create unprecedented political challenges for the governments, corporations and people on whose behalf laws are created and administered. These political and policy-making challenges currently cripple governments and other international actors.

Unless or until all states can develop new sources and forms of energy production that do not cause environmental damage, the impacts of curbing the emissions outputs and their effects will require everyday people to accept compromised lifestyles (de Coninck et al., 2007). The comfort of their housing; ease of transport and mobility; quality of services and infrastructure; and variety and quality of foods will all be affected by these changed political and economic conditions. These political challenges will gain increasing intensity in coming years as people and governments confront a greater range of problems in maintaining their current production, consumption and lifestyle practices, all of which rely upon access to secure and affordable energy sources (United Nations Development Programme, 2008). This reality is part of the reason for many people and governments continuing to cling to hopes of technological solutions, including their beliefs that new plentiful cleaner energy supplies are available.

It is precisely '[b]ecause of the diversity of energy's roles in society' that many government leaders struggle to respond to 'the multiplicity of energy-related challenges' they face (Anadon and Holdren, 2009, p. 98). In spite of the hopeful anticipation of technologists many, including governments, economists, industrialists, farmers and workers of many kinds are still coming to terms with 'the fact that no energy technology currently known or imagined has the versatility and other characteristics needed to address all facets of the energy situation at once' (Anadon and Holdren, 2009, p. 98). The diverse uses of fossil fuels and other established sources of electricity have afforded forms and rates of consumption that are unable to be supported by new energy sources. Significant political obstacles lie in the changed attitudes and expectations associated with accepting these new political and economic realities (Eakin et al., 2009). In so far as new energy sources have been identified and new technologies have been developed, presently these remain quite specific in functional applications. Many of these new energies are also limited in their functional applications or adoption because of other climate change–related issues (Marland and Obersteiner, 2008).

Perhaps the greatest of the challenges presented by 21st-century energy and climate change problems is the overturning of our beliefs in progress. Successive generations of people have now benefited from new technologies, new forms of production, and new economic opportunities supported by ready sources of energy (Page, 2006; Flannery, 2005). A challenging reality of energy–climate change, is the fact that ' "there is no silver bullet" – no single new or improved technology on which hopes for answers to the big energy questions can or should be pinned' (Anadon and Holdren, 2009, p. 98).

New energy policies

Energy debates present additional challenges because sustainability relies upon the widespread adoption of alternative sources and production of energy. The level of change required is immense, and states with established energy infrastructure are surprisingly resistant to change (de Coninck et al., 2007; Johansson and Azar, 2003). Given the reliance of states and private economic actors upon innovation in triggering economic growth and securing their profitable market share, one might have anticipated broader willingness and ready acceptance towards changed production practices. Industrial capitalism has revealed itself to be quite reluctant to change the means or modes of production, preferring the achievement of incremental gains in efficiency over wholesale changed practices. This is partly a consequence of the levels of infrastructure and capital investment involved in industrial production and the nature of political systems and human association more generally, which seek to create stability and order, and thereby demonstrate resistance to structural change (Bazerman, 2009; Hogan, 2009).

The challenges of climate management and energy policy extend to the expectations held by individuals and social groups as well as economic actors, including banks, insurance companies, agricultural producers, importers and exporters (McCartney and Hanlon, 2008; Manne and Richels, 1995). They extend to apparently trivial issues such as the availability of plastic shopping bags at supermarkets, whether or not meat is packaged on polystyrene trays with plastic backed absorbent material as lining, and then covered with cling wrap. They extend to whether or not energy-saving light globes are compulsory. They extend to the ways that modern prisons and hospitals function, and how well-lit city streets are maintained to ensure public safety.

Energy policy affects fuel supply and pricing, determining the types of fuel, electricity and gas that societies utilise, including apparently mundane matters, such as packaging, and important, but not terribly visible, matters such as running life support machines and incubators for fragile newborn infants. Addressing climate change through energy policy is a necessity across the world. The effects of energy policy decisions will trickle down into everyday life and produce consequences for personal security, lifestyle convenience and the protection of vulnerable lives in myriad ways that are currently difficult to imagine. While the decisions and actions to combat global climate change will be undertaken at the highest levels of national and international governance – both state and global – their effects will be most sharply experienced at the individual level (Sawin, 2003).

Generating alternatives

Within international climate change and energy debates, discussions include considerations of the potential of a variety of alternative, non-fossil fuel and renewable energy resources (Morse and Myers, 2001; Pasternak, 2000). Some of these alternative potential energy sources require futuristic and substantive new technological developments to become alternative sources of electricity or heating, or as partial solutions to reducing global greenhouse gas emissions. Others require new or different forms of infrastructure to achieve broader adoption or greater commercial viability although they arise from previously established energy sources. This range of alternative potential energy sources serves to cloud the already complex sets of issues and debates concerning energy and climate change responses.

While a general level of optimism concerning the capacities of new technologies and alternative energies to support comfortable lifestyles seems appropriate, it also seems wilfully naive to presume that these can, or should, support current levels of affluence and high rates of resource consumption (Turton and Barreto, 2006; Sawin, 2003). The switch to cleaner energy sources among developed states will require significant political will and public expenditure. Shifting away from brown coal as the major source of electricity generation, for instance, will require new infrastructure as well as new power plants. States will need to create and support private actors seeking to establish viable alternative energy supplies. Achieving these goals will rely upon changed relations between government policy-makers and other economic actors as they seek to achieve new energy use and distribution targets.

The levels of adaptation strategies required for switching societies to renewable and alternative energy sources will be enormous. These energy challenges cannot be reduced to questions of how we light our homes and offices because they also relate to how we produce our food, and how we transport the goods and commodities that comprise our markets. While a shift away from energy consuming plastic shopping bags provides an example of a relatively easily implemented policy, the question of what provides an optimal alternative means of enabling shoppers to carry home their purchases also illustrates the complexities of establishing appropriate and responsible new preferences. For instance, it is not necessarily less energy intensive to produce reusable shopping bags, or to produce biodegradable packaging. Both approaches must take into account the locations and types of production inputs that are required and the distribution networks associated with the products.

Industrialised societies have grown accustomed to the regular appearances of new technologies, and their economic histories show that these are more readily adopted as items of consumption or pursued as refined means to more profitable production (rather than changed production processes). Hence, many of the energy or resource conserving 'technologies' or products recently introduced to markets in developed states represent refinements of old technologies, such as dual flush toilet systems, solar heating, heat pump hot water systems and electric powered vehicles. These evolved patterns might well be challenged as international efforts to fast track new technologies and commercially available alternative energy sources disrupt current and familiar market-based practices.

Conclusion

Establishing international climate and energy policies that support sustainable and secure societies in the 21st century will require the international political community to respond to deeply challenging political complexities. Developing sustainable lifestyles and forms of human association and economic production in the face of these challenges will require more government intervention, not less. Reliance upon market-driven pricing, investment and commodities production has contributed to an over-reliance on the use of fossil fuel energy sources, especially among industrialised societies and their economic sectors (de Coninck et al., 2008, p. 337). Many developed states have demonstrated an over reliance upon markets to determine costs, thereby enabling private economic actors to play key roles in determining the energy policies of their states. These dynamics are likely to be subject to dramatic disruption as states collectively seek to reassert their political authority.

Interplay between the climate consequences of current energy sources and their uses would be problematic for the international political community even if all of the world's states were more evenly developed, held more equitable resource bases and the distribution of wealth between them was more just. The absence of equity in their economic circumstances creates additional complexities for the establishment of new energy systems. Intractable difficulties associated with global injustices in economic development and wealth distribution form part of the political context for international climate change responses, especially in relation to energy resource production and consumption debates. For these reasons, energy and climate debates especially challenge the capacities of states and the international political community to maintain their dreams of continuing economic prosperity, material progress and expanding rights arising from democratic political visions.

7
Energy and the Security Dilemma

Introduction

At present, it is difficult to imagine a future when the current reliance upon fossil fuel energy sources are not central to the activities of human societies, their economies and the efforts of governments and other actors to maintain social order. This is especially true for developed states where higher energy requirements support daily life and the extensive networks that maintain social order. The nature of developed states and the affluent lifestyles they support rely upon high rates of energy consumption and expectations of their continued availability such that an 'abundance of energy is what defines life in industrial nations and distinguishes it from traditional' societies and lifestyles (Prugh et al., 2008, p. 101). Consequently, it has become important to confront global climate change while also finding and developing sources of energy that are less environmentally harmful. As has been observed by Bill Gates (cited in Mosher, 2011, http://www.wired.com/business/2011/05/bill-gates-energy-tech/) 'if we don't have innovation in energy, we don't have much at all'.

Debates about the urgent need to find more sustainable energy sources for the future have been around since the late 19th century but came to prominence more recently during the 20th century with alarms that oil supplies had already peaked (Armaroli and Balzani, 2007). Global climate change brings peak oil debates and the need for viable alternative energy sources into sharp and immediate focus. Most scientists agree the world retains sufficient fossil fuel reserves (oil, gas and coal) to sustain ongoing industrial growth. However, the level of growth that might be sustained is questionable and the environmental damage arising from greenhouse gases involved in using fossil fuels make their use dangerous and unsustainable. While the 'world has slightly over 1 trillion tons of coal – enough for more than 200 years at present consumption rates... using more presents serious problems' (Mason, 2003, p. 12). If states are to pursue further industrial development they need to shift to alternative combinations of energy sources and the timeframes for doing so appear to be calamitously short and will result in substantial insecurity and conflict unless carefully managed.

The need to transition to new energy sources makes addressing and mitigating global climate change an especially difficult process. Developing cleaner energy sources involves decisions about which will prove cost effective and sustainable into the long-term future, and how quickly the infrastructure to enable their ready distribution might be constructed. Many of these choices involve factors and variables that are uncertain and often the costs of developing new sources of energy remain unclear. These circumstances are challenging for governments and investors who prefer economic certainties. Uncertainties about technological viability, supply, cost, effective uses of available resources and the timeframes for their development all confront states with insecurities that may impact upon their internal and external capacities. Balance of power considerations are never far removed from deliberations about national interests and these have routinely been prioritised above international and environmental imperatives (O'Neill, 2009).

States face major political challenges in reconciling policies to secure long-term energy supplies alongside economic, industry and urban politics. They face even greater political challenges in reconciling these policies while supporting the economic aspirations of developing states. The need to do so while simultaneously addressing global climate change increases the intensity of these challenges. As noted in Chapter 5, population growth and technological advances in developed and developing states alike have intensified global energy use. The global rate of resource use has risen from 70 per cent of annual global capacity in 1961 to 120 per cent in 1999, making plain that such resource use cannot be sustained by the earth's biosphere (Wackernagel et al., 2002).

The transition to a new energy paradigm will be fraught with uncertainty and conflicting agendas, aspirations and national interests. As observed by Pielke (cited in Bazilian et al., 2011, p. 3751):

> When GDP growth comes into conflict with emissions reduction goals, it is not going to be growth that is scaled back ... when rich countries wanting emissions reductions run into poorer countries wanting energy, it is not going to be rich countries who get their

way. When energy access depends upon cheap energy, arguments to increase energy costs or deny energy access are not going to be very compelling.

This view has been reiterated by political leaders, including 'Russian President Vladimir Putin [who] has said that the order of business should be "first the economy, then the environment" ' (Bell, 2006, p. 111). Further complexity arises from the reality that '[i]n a world of increasing independencies, energy security will depend much on how countries manage their relations with one another' (Yergin, 2006, p. 82). Ensuring energy security while addressing global climate change will then rely upon politics, international and domestic, rather than new technologies.

As the international community develops climate change governance mechanisms it must also reconcile the security implications and tensions concerning the use, costs and distribution of resources. Many international discussions concerning energy have experienced 'one step forward, two steps back' trajectories for a number of reasons. First, the energy that is produced from fossil fuels are significant sources of greenhouse emissions, and their use might be expected to become increasingly regulated in the near future. Second, these are insecure energy sources as many believe their supplies to be dwindling in readily accessible locations or supply levels (de Almeida and Silva, 2009). Predictions of an impending 'fuel drought' that might 'cripple world trade and economies' have become more common in recent years (Mason, 2003, p. 9). Third, oil and natural gas resources are predominantly located in either politically sensitive and potentially unstable locations, such as the Middle East, or environmentally unstable and sensitive locations such as equatorial Latin America and Russia. Each of these factors contributes to an emerging international appreciation that 'a sound security of supply concept call for, among others, a diversification of technologies and energy sources' (Turton and Barreto, 2006, p. 2233).

Complexity and complications

Energy-climate change debates are complicated by realisations that '[s]olving the climate change problem requires a wholesale change of technology, particularly in the energy and transport sectors' across the world (Alfsen et al., 2010, p. 210). Many states face increasing energy demands and policy decisions about which sources might be secure into the future and affordable for most of their industries and people.

For developing states these tensions compound as they simultaneously confront increasing energy needs, growing populations and increasing urbanisation. Crucially, 'most of the additional energy' required to support a growing population and rapidly expanding cities 'is projected to come from fossil fuels, [and]...meeting these demands with conventional fuels and technologies will further threaten the natural environment, public health and welfare, and international stability' (Sawin, 2003, p. 85).

For many developing states, the problems of securing access to energy resources, and the financial capacities to develop infrastructure and technologies, underpin their abilities to imagine a viable future for their peoples. Affordability has been an important determinant of expanded access to electricity that increased industrial production and modern lifestyles have relied upon (Ahuja and Tatsutani, 2009). The international political community now recognises that '[r]ising energy costs could lead to a global economic recession and compromise many of the systems which make the modern way of life possible' (McCartney and Hanlon, 2008, p. 655). Addressing energy security, while simultaneously confronting global climate change, is a significant challenge for every member of the international community. This political reality is partly a consequence of the distribution and availability of energy resources, most notably fossil fuels, the environmental consequences of using these and the relative costs of cleaner energy sources.

In some respects, developing and developed states face different sets of issues in resetting their policy priorities to meet the energy-related challenges of climate change. Different economies and forms of political organisation influence their abilities to imagine alternative futures and develop new strategies for preserving their societies. International security concerns arise from familiar observations about unequally distributed contributions to greenhouse gas emissions. Those who use the most energy emit the greatest volumes of greenhouse gases, and these people reside within developed states (Flannery, 2005). Those who most frequently experience fuel resource and energy-related shortages, and are most vulnerable to rapid increases in energy pricing, reside in developing states. These issues are further complicated because the bulk of petroleum- and oil-related products are produced by developing states but consumed by developed states. These observable patterns are made exceptionally politically sensitive by global climate change and considerations about who should ethically bear the costs of securing the future (Page, 2006, pp. 167–173).

For many developing states, economic growth presents an 'overriding policy priority' in the pursuit of prosperous lifestyles associated with, and afforded by, developed economies (United Nations Committee for Development Policy, 2009, p. 9). However, as investors and political actors seek to secure their well-being they become increasingly sensitive to the fact that '[e]nergy is embodied in any type of goods and is needed to produce any kind of service. This is the reason why it takes energy to improve people's standard of living' (Armaroli and Balzani, 2007, p. 54). Satisfying these aspirations will only be possible if 'the reach of energy and infrastructure' expands to become 'available to a larger proportion of the population' (United Nations Committee for Development Policy, 2009, p. 9). This is unlikely to be possible in the 21st century as the impacts of climate change alter the cost–benefit equations of utilising current forms of energy production and energy demands continue to increase to satisfy a growing global population (Stirling, 2007; Sawin, 2003).

Energy issues are a central component of global security and industrial vulnerabilities for all states and the societies they create and maintain. As observed by Rhodes and Beller (2000, cited in Pasternak, 2000, p. 2):

> Development depends on energy, and the alternative to development is suffering: poverty, disease, and death. Such conditions create instability and the potential for widespread violence. National security therefore requires developed nations to help increase energy production in their more populous developing counterparts.

A globalised economy and the aspirations of people in both developed and developing states are underpinned by expectations of affordable, accessible and secure energy supplies.

Since the middle of the 20th century, 'economic growth and health have grown in parallel. Greater wealth leads to better health, but improved health also contributes to economic growth' (McCartney and Hanlon, 2008, p. 653). Health considerations demonstrate how energy debates can become very personal and highlight complex ethical imperatives underpinning new energy futures. Issues that impact upon climate change mitigation and adaptation approaches are linked with other considerations including the roles of private economic actors, debates about the relative merits of various energy sources, their production and distribution mechanisms, and when oil supplies will become unviable. In the 19th and 20th centuries, developed states established their economies and societies through intensive use of energy (Ponting, 1991, pp. 267–294).

Additional complexities for developed states arise from deep economic vulnerabilities associated with recent global financial crises and an appreciation of the risks associated with reliance upon concentrated energy sources (Turton and Barreto, 2006). As argued in Chapter 9, expectations of progress and aspirations of increasing affluence have been important influences upon political decision-making and the expansion of international markets over the last several hundred years. Current modes of energy production and consumption are major contributors to global carbon dioxide and other greenhouse gas emissions and high levels of water consumption. All states face risks to their abilities to secure continuing economic growth and provide social order into the future.

An obvious difficulty in responding to global climate change arises from conflicting priorities: those who seek to establish greenhouse emissions targets also seek to maintain economic prosperity bearing in mind that all decisions, political actions and productive activities 'have consequences, both in the near term and into the future' (Hammitt, 2000, p. 388). The extent to which these priorities are understood as conflicting then forms the basis of debate about what can or should be done and at what cost (Mulligan, 2010; McCartney and Hanlon, 2008; Pasternak, 2000). As noted by Armaroli and Balzani (2007, p. 52), 'opportunities discovered and exploited by one generation can cause challenges for subsequent ones'. Achieving energy security and avoiding the worst of the projected climate change scenarios means overcoming 'the major obstacle' of ensuring that 'international affairs, politics, decision-making by governments, and energy investment and new technological development' support policy goals (Yergin, 2006, p. 75).

States' abilities to influence others and pursue their strategic goals depend upon secure access to energy sources that are necessary to support production and the lifestyles expected by their peoples (Barnett, 2001). As Hammitt (2000, p. 389) observes, 'shifts of industry and agriculture among nations, and accompanying migration of populations can be highly disruptive. Political barriers to migration and the costs of developing new infrastructure may severely limit the extent to which adverse impacts can be reduced.' It is impossible for the political debates concerning energy production, use and consumption to be separated from aspirational lifestyle and social considerations because states' abilities to exercise authority and pursue goals of economic growth and social harmony will be challenged in the next decades as global climate change unfolds (Barnett, 2001).

Unequal contributions and risks

Conducting meaningful discussions concerning regulating the use of energy resources means paying attention to questions of global inequalities. Patterns of energy production and consumption are aligned with patterns of relative affluence, poverty and observed distributions of greenhouse emissions. Climate change mitigation and abatement policies affect the future energy security of states and are linked with agendas to more ethically redress present inequalities in energy access and use (Gardiner and Hartzell-Nichols, 2012; Armaroli and Balzani, 2007; Page, 2006). While this is necessary in order to ensure the participation of developing states, the tensions it creates within the international community provide political complexities that can delay progress.

As the global population increases, demand for energy supplies increase disproportionately especially among developing states where current demands for energy are comparatively low (Sawin, 2003). The magnitude of anticipated energy demands are such that '[p]er capita energy consumption in the developing world is expected to increase fourfold to sixfold over the next century' (United Nations Committee for Development Policy, 2009, p. 15). Such increases in demand are unlikely to be addressed by electorally unpopular policies intended to replace current lifestyle aspirations with more modest standards. Under these conditions, all states will experience difficulties in coordinating their short-term roles and responsibilities for protecting their territories and people with the longer term pursuit of more sustainable energy production and consumption practices. Secure energy supplies are essential to the policies, regulations and legislation that states create to ensure social harmony, public order and prosperity. Maintaining access to secure energy supplies has obvious importance for economic activities, but in modern societies' energy also matters for the movements of people and goods, housing, medical care, and population management initiatives as well as access to water and food supplies (Roaf et al., 2005). Predictable and affordable energy supplies are fundamental to states' interests and to their efforts to consolidate their roles within the international political economy and community (Barnett, 2001). For this reason, energy diversification is often perceived as a vital component of future energy security.

Security and order

Energy insecurity has long been acknowledged as a potential threat to order within and between societies (Sawin, 2003, p. 88). Ensuring energy security is a fundamental requirement of both domestic and international order, and it will become increasingly complicated as the effects of global climate change are more broadly experienced. Internationally imposed restrictions on the use or production of certain energy sources, and altered trade and economic structures, will impact upon the abilities of governments to provide social order for their peoples (Bazilian et al., 2011). As governments and other international actors endeavour to establish a series of mitigation and adaptation strategies to deal with global climate change, they must also collectively engage with questions of how to regulate energy use to minimise risks of disorder arising from energy insecurities.

As the consequences of global climate change begin to be felt more widely and frequently through the 21st century and impact upon energy security through severe weather disruptions to infrastructure they will alter prospects of economic growth (Umbach, 2010; Yergin, 2006). In particular, global 'consumption of electricity, the most versatile form of energy, will increase even more sharply by most estimates – nearly 70 per cent' (Sawin, 2003, p. 85). It is hard to imagine that this growth will occur without significant greenhouse gas emissions and other environmentally damaging impacts. While the moral claims of developing states to extend their economies and increase prosperity among their people are thoroughly legitimate, these aspirations must now be located in the context of other responsibilities, such as preserving an environment that can support human communities into the future (Arnold, 2011; Gardiner, 2011).

The imperative of addressing energy supply demands while also taking action on global climate change will challenge international order and security. The forms of government that provide the greatest rights and opportunities to their people rely upon complex energy intensive networks between their social and political actors. By contrast, repressive forms of government utilise direct forms of coercion in their efforts to maintain order and these are more common in societies that currently lack high levels of industrial development and the economic prosperity associated with it. The liberal project of international democratisation relied upon continued industrialisation and unfettered access to affordable energy. Policies that curtail the availability of affordable energy, in order to address global climate change for the benefit of future generations, may do so at the expense of contemporary societies. Obviously this is unpalatable to current political and business leaders whose decision-making and careers are dominated by immediate timeframes. For different reasons, concerns about long-term environmental sustainability are also often overshadowed by more immediate concerns about energy access and affordability among developing states (Ahuja and Tatsutani, 2009, p. 2).

Contemporary governments and political leaders are concerned with climate change and energy security because they are aware that the political institutions and relationships that characterise modern societies cannot be sustained without secure and affordable energy sources. In August and September 2003, a series of large-scale electricity failures occurred in Italy, Denmark, Sweden, Malaysia, the United States and Canada, affecting millions of people in each instance, shutting down industries and leaving major cities without electricity (Andersson et al., 2005; Roaf et al., 2005). These 'demonstrated clearly that many countries have become absolutely dependent, for the ordinary functioning of society, on a constant, and high quality, supply of electricity' (Roaf et al., 2005, p. 298). They highlighted the vulnerability of industrialised states to electricity shortages, and point to the challenges they face in maintaining secure electricity supplies in the 21st century and reconciling the competing imperatives of energy security, economic prosperity, social order and sustainable progress.

Global energy production and consumption will change as the uses of fossil fuels become increasingly regulated through international efforts to address global climate change. How states choose to satisfy their energy needs within their own borders is becoming a concern of other states and the international community (Bazilian et al., 2011, pp. 3751–3752). Already cleaner energy technologies are being adopted and renewable energy sources utilised in both developed and developing states. Nonetheless, as global climate change unfolds in the 21st century, many states face prospects of increased insecurity in relation to the ready availability of affordable energy supplies. As McCartney and Hanlon (2008, p. 655) point out, these political challenges will intensify as more people realise 'the question is not whether global temperatures and oil prices will rise, but by how much and how soon. Therefore, the most sturdy nail in the coffin of economic growth is that of its unsustainability.'

Although the contemporary economic and political challenges of ensuring energy security are well understood, governments and policymakers are only just beginning to consider the depth and breadth of these challenges for their future choices. Energy security challenges, such as the needs for new infrastructure, cleaner and diverse sources of energy and growing demands for electricity, will affect future political behaviour in ways that extend beyond self-contained energy policies or efforts to regulate energy production and consumption (United Nations Committee for Development Policy, 2009, p. 9; Yergin, 2006). One of the reasons we can be so sure of this changing international political context is because there are simply not the resources available for the entire global population to live at the levels presently enjoyed in developed states (Armaroli and Balzani, 2007, p. 63). As a consequence, states could be expected to begin to pursue the security of future energy needs more aggressively, or to use more of their energy resources to guarantee their own economic and development agendas.

'[R]ecent trends in renationalisation of energy policies and concomitant resource nationalism' mean that '[s]tate owned energy companies now control far more oil and gas reserves (up to 85%) than... traditional private energy companies' (Umbach, 2010, p. 1232). There are both benefits and disadvantages to these developments. One benefit is that state ownership or control over resources and their use might result in coordinated synergies between policies of energy security and climate change mitigation and adaption (Turton and Barreto, 2006). A second benefit is that the global energy supply chains tend to lie within the regulatory ambits of governments whose concerns and responsibilities are broader than the 'business interests of companies' which are more commonly 'guided by short-term economic benefits' (Umbach, 2010, p. 1229). Private companies are especially prone to protecting their short-term economic goals when they face 'an increasingly competitive environment' such as that presented by declining energy and other resource security (Umbach, 2010, p. 1229).

Chief among the disadvantages of government ownership of energy resources are that in an increasingly globalised and interconnected energy market resource competition between state entities (rather than private corporations) may be more likely to lead to the use of resources and infrastructure (such as transnational pipelines) as foreign policy weapons. Concerns that efficiency, and thus energy affordability, may be at risk because 'privately held energy corporations are more efficient and productive organizations than any government-held entities' may be less relevant (Umbach, 2010, p. 1239). Market efficiencies are driven by short-term interest in profits rather than long-term interests in sustainability. If the security implications can be managed and potential conflicts avoided then greater, direct involvement by governments in the energy supply chain may well prove beneficial over the long term.

Resource insecurity will not be limited to meeting projected energy needs, and the range of actors seeking to secure access to energy sources will not be limited to states. The economic importance of energy resources is such that they are likely to evoke intense reactions among a range of actors who are willing to go to some length to secure their supply into the future. Most governments recognise that their own decisions and actions are unable to protect them from climate change but are familiar with taking steps to ensure their own security although they risk being perceived as threatening by other states (Edmondson and Levy, 2008, p. 17). Governments that are concerned with their own security can only assess this within the broader context of the vulnerabilities and changing interests of others. This dynamic affects more than military security and defence policies as it creates patterns of trust among allies and distrust among potential enemies and others whose shared interests become clouded by perceived threats.

It is easy to imagine conflicts will occur between states concerning their opportunities to extract, exploit and produce energy to meet the continuing and increasing demands of their populations. However, this is only one potential trigger because conflicts might also occur over climate change induced food shortages or redistributions in water resources (Evans, 2010; Conway, 2009; Peters, 2003). As noted by Pasternak (2000, p. 2), '[c]are must be taken with energy development, as with all development, to maintain environmental sustainability. The roots of conflict may also be found in scarcity that results from environmental degradation.' Concentrations of key resources and energy production sites also make many states vulnerable to acts of terrorism which are likely to increase as energy shortages are more widely experienced. States might experience further security risks arising from failures to meet internationally agreed greenhouse gas emission targets, or to curb their use and production of dirty energy sources. States that jeopardise the survival and prosperity of others by continuing to use brown coal for electricity production with relatively rigid grid structures, might find themselves subject to greater international intervention. Similarly, states that do not readily share their resources are also likely to experience international intervention to pressure them to amend their practices.

Cleaner energy

Satisfying future energy requirements and addressing climate change will require a global shift away from current levels of reliance on fossil fuels. Faith in human ingenuity that produced the modern world provides optimism about the potential of developing cleaner energy solutions as a means of responding to global climate change (Stern, 2009). The hopes underpinning such optimism focus on cleaner energy sources in reducing greenhouse gas emissions while sustaining industrial production and affluent lifestyles without particular regard to broader political issues, such as the gaps between developed and developing states, or competing demands for limited resources, such as water and forests.

Despite reservations about the ease with which such technologies might be identified, the timeliness of their development and the degree of rationality necessary for their global adoption there is no denying they will play a significant role in addressing climate change (see Chapter 3). Globally, the transition to cleaner energy sources will be an enormous undertaking requiring a coordinated shift from well-established and well-understood energy production chains in fossil fuels to others that are less well understood. For many states, uncertainty about the economic advantages and disadvantages of cleaner energy sources remain a significant impediment to their global adoption and will delay shifts away from the immediate security of fossil fuels. Biomass 'production often demands more energy than it yields' since it often involves capital intensive grand-scale agricultural production which is dependent on petroleum by-products as well as fossil fuel driven cultivation and harvesting methods (Mason, 2003, p. 16). Even ardent green advocates are concerned that 'carbon emissions over the lifecycle of the production chain for some biofuels may not be significantly less than the carbon emissions from gasoline production' (Lee, 2009, p. 75).

Debates about the advantages and disadvantages of alternative energy sources will intensify into the future as states begin to diversify their sources of energy supply and the advocates of different energy sectors and associated industries advocate for government investment. The privatisation and commercialisation of energy sectors, and the admission of NGO voices into various forums of the international community, have publicly politicised this realm of government policy. While 'we cannot rely on what is presently unknown' (Armaroli and Balzani, 2007, p. 64) in terms of alternative energy sources, many of the renewables are not new, have been around for decades, and are in principle well understood. Their application, however, has been stymied by a comfortable reliance upon established energy sources and expectations of 'creat[ing] a new era of progress and prosperity...discover[ing] new technologies and sources of energy [to]...make our energy supplies more secure' (Stern, 2009, p. 4). Such views have, to date, justified policies of inaction as political, social and business leaders continue to seek new technologies that will resolve current problems without compromising political and economic aspirations. More sober commentators have suggested:

> Such illusory behaviours result from a false belief in our control over the most uncontrollable of events. In the realm of climate change, this type of positive illusion is represented in the common expectation that scientists will invent technologies to solve the problem. Unfortunately, there is little concrete evidence that new technologies will solve the problem in time.
> (Bazerman, 2009, p. 167)

Fortunately, some of the cleaner energy sources that are already well known provide foundations from which to respond without relying upon new technologies, even though these will always be welcome.

Efforts to support continuing industrialised lifestyles, and promote new levels of industrialisation and prosperity among developing states, are likely to rely upon sophisticated agreements, infrastructure, and sustained cooperation between governments, international organisations and economic actors to ensure the provision of 'a stable and reliable supply of energy' (Mason, 2003, p. 17). Without secure and reliable energy supplies, agreements may lapse as political leaders prioritise the short-term interests of their own political constituencies over other considerations. Achieving the goal of sustainable, secure, energy supplies will rely upon cooperative endeavours across political and economic sectors, and across geographic areas governed by different political actors and structures. Achieving this goal will also require the use of multiple new cleaner energy sources. For instance, it is likely to require energy sources to be 'used in combination – utilizing water, wave, tidal, wind and solar power as appropriate' to meet the energy needs of 21st century societies (Mason, 2003, p. 17). Such diversification is to ensure security of supply and protection against acts of terrorism or extreme weather events while accommodating the current technological limitations of cleaner energy sources (Turton and Barreto, 2006; Yergin, 2006).

126 *Climate Change and Order*

Many of the debates concerning the potential uses of cleaner and renewable energy sources in the 21st century express a mix of concerns regarding the need for new technologies and new patterns of technological distributions across the world. To some extent, these debates centre upon questions concerning 'the most effective policies and institutions for achieving the dramatic changes and associated reductions' required to stabilise emissions and begin to address global climate change impacts (de Coninck et al., 2007, p. 335). For these debates to become effective sites of negotiated agreements about future patterns and practices of energy production and consumption, states, international organisations and key economic actors will need to take account of the:

> significant differences between fossil fuel sources of energy and renewable alternatives in terms of both cost and convenience. Shifting to cleaner alternatives without jeopardizing development or equity goals would be impossible for even the most dynamic economies without external financial support and technological assistance.
> (United Nations Committee for Development Policy, 2009, p. 17)

Hence, new institutional arrangements and structures will be crucial to international climate-energy responses, in terms of achieving more equitable access to, and distributions of, resources and supporting mitigation and adaptation technologies.

A new energy paradigm

For several decades, scholars and policy-makers have argued that the responsibilities of contemporary states in the near future will, in all likelihood, include a willingness to recognise global priorities ahead of local domestic goals, including economic growth. However, shifting this idea beyond rhetorical exchanges is problematic because it involves states placing collective security considerations ahead of domestic security and prioritising social goals ahead of economics. These dynamics are highly problematic for states whose abilities to rapidly remodel themselves are limited by their abilities to evolve long established and commonly held features of sovereign statehood. Optimism about international responsiveness to impending political and security and energy crises might be ill-placed, especially if non-state actors are not included in new initiatives. The debates surrounding these issues are magnified by pre-existing tensions arising from long-standing global economic and political rivalries and inequalities. These tensions will increase as states experience increasing resource insecurity in the 21st century and confront the effects of global climate change.

It has been argued that the relationship between human beings and their natural environments might be altered as human communities come to terms with the realisation that high rates of energy consumption threaten their continued existence (Eckersley, 2004; see also Dryzek and Schlossberg, 2005). While the philosophical and ethical underpinnings of these debates lie outside the scope of this chapter, they remain important influences upon the political contexts within which global climate change and energy debates occur. Permitting present levels of greenhouse gas emissions affords economic and political advantages in avoiding major disruptions to current industrial activities and established political structures but doing so 'may increase the probability and severity of climate change and the costs of limiting it' (Hammitt, 2000, p. 388). On the other hand, 'reducing emissions, through energy conservation or substitution of alternative sources, may... impose other risks', such as increased volumes of 'nuclear wastes or habitat destruction from hydropower' (Hammitt, 2000, p. 388). There are rarely easy choices in effecting mitigation and adaptation responses to global climate change, in energy policy decision-making, tensions and contradictory impacts are especially fraught.

States have thus far avoided establishing firm greenhouse gas emissions reductions targets and international regulation of energy markets in order to avoid shouldering the economic and political burdens of exercising decisive political leadership. By waiting to see what others will do they have hedged political risks against economic choices. 'Hedging' and the precautionary principle both shape international policy decisions. Hedging occurs when decision-makers seek to minimise the potential risks in existing investments, such as in maintaining or justifying the future viability of current technologies, or protecting their assets, such as in advocating particular forms of energy that might benefit their competitiveness in energy markets (Yohe et al., 2004; Manne and Richels, 1995). By contrast, the precautionary principle seeks to incorporate ethical considerations into risk management assessments.

In global climate change discussions, advocates of the precautionary principle argue that responsible decision-making requires leaders to prioritise the protection of people and societies over other considerations (United Nations Educational, Scientific and Cultural Organization, 2005b). Adopting a predisposition towards hedging can often be equated with the pursuit of 'wait-and-see' approaches to climate change impacts, while the precautionary principle is more likely to result in 'anticipatory measures' to minimise future risks (United Nations Educational, Scientific and Cultural Organization, 2005b). Both are important for understanding the political dynamics that characterise international debates concerning sustainable energy, and each can contribute to effective decision-making. However, as Sawin (2003, p. 105) points out, '[g]overnments need to eliminate inappropriate, inconsistent, and inadequate policies that favour conventional fuels and technologies and that fail to recognize the social, environmental, and economic advantages of renewable energy'.

It is easy to acknowledge that all contemporary governments 'face pressing economic, environmental, and security challenges arising from the energy sector' but this does not give them new powers or abilities to reconcile their ongoing needs for electoral support, economic growth and security (Anadon and Holdren, 2009, p. 1). These dynamics lead to policies and agreements that are limited in their potential effectiveness, contradictory or confused. Analysts such as Hammitt (2000) believe that incorporating the precautionary principle into international climate change discussions could assist policy-makers in achieving new decisions in the face of continued uncertainties. For instance, Hammitt (2000, p. 397) argues that the 'precautionary principle would appear to be particularly well suited to environmental threats such as global climate change, which are characterised by a high degree of uncertainty about future consequences, high inertia which prevents expeditious correction of adverse trends, and the possibility of catastrophic outcomes'. This does not mean that established models of decision-making by political leaders will or should be rejected out of hand, or that the model afforded by the precautionary principle will always lead governments towards effective decisions.

Democratic decision-making within the international political community is a further complication in the security risks arising from the interplay between energy-related production and consumption issues and the creation of policies for responding to global climate change. Throughout the 20th century, international decision-making among states tended to occur through broad levels of collective participation and agreement over terms, decision-making processes and responsibilities for implementation. International security agreements and political structures typify these approaches. A generalised belief in the value of the rights of sovereign states to participate in broadly democratic policy forums characterised the establishment of key international political mechanisms through the 20th century. The dominance in political leadership by liberal democratic states produced international political structures that favoured democratic processes.

Unfortunately, the protracted timeframes required for enabling democratic decision-making among the diverse membership of the international political community make it unlikely the complexities of energy-climate change can be addressed in a timely manner. As observed by Hogan (2009, p. 128):

> Initiatives to improve energy security...and transform the structure of energy systems all anticipate major infrastructure investment. Long lead times and critical mass requirements for these investments present chicken-and-egg dilemmas. Without the necessary infrastructure investment, energy policy cannot take effect. And without sound policy, the right infrastructure will not appear. Acting in time to provide workable policies of infrastructure investment requires a framework for decision-making that identifies who decides and how choices should be made.

The United Nations Law of the Sea took 24 years to implement and ratification of the Kyoto Protocol required more than seven years. In stark contrast, consensus within the scientific community is that climate change responses are now urgent. At a minimum, it is now considered imperative to establish greenhouse gas emissions reductions in excess of those previously agreed to limit global warming to 2°C.

The overarching principles of consensus-based management and collective security have been entrenched in the international community through inclusive membership with relatively equal rights of participation in international political institutions. Notwithstanding the dominance of liberal ideologies and capitalist economics as central to the activities, dynamics and structures of the international political community, the presence of democratic decision-making has enabled broadly inclusive decision-making practices. These have been especially important in ensuring that developing and less developed states have enjoyed rights to international political participation. In recent decades, principles of mutual defence have been extended to support humanitarian interests and consolidated in peacekeeping and security structures. The international political community now perceives states as responsible for ensuring the survival of people across state boundaries. Hence, the security of people has become an important component of international security measures, where it sits alongside earlier collective security principles which had granted and established such mutual responsibilities in relation to protecting the interests of, and preserving, sovereign authorities.

Through collective security, states agree to protect each other from acts of unwarranted, unjustified, intervention by other states. As a result, states rely, to greater or lesser extents upon the support of the international political community in order to meet their security interests. In doing so, they overcome some of their individual limitations in assessing and managing security risks. One of the consequences of collective security having become widely accepted within the international political system has been to increase the effectiveness of international diplomacy. This does not mean that collective security leads states to experience diminished levels of domestic responsibility for managing and defending their people, territories and resources. Rather, it emphasises their responsibilities for ensuring that borders are not breached, and illustrates how states now also hold additional responsibilities for protecting others.

In the past, the international community has established collective governance mechanisms for problems that have been shared among states, or when the actions of a few have been perceived to risk the security of many. The regulation of nuclear and other weapons of mass destruction and transboundary pollution provide notable examples. Managing the security dimensions associated with energy production and consumption, and developing climate change mitigation and adaptation policies, present such complex issues that international regulatory efforts have repeatedly stalled, or fallen short of their goals.

Conclusion

In responding to energy use–climate change dynamics and related security issues, it will be politically important for all states to establish and contribute to meeting emissions targets. Those who seem unwilling to participate in international agreements or refuse to establish meaningful and attainable targets may find their abilities to participate in international markets, and to enjoy the protections afforded by collective security arrangements, severely compromised. Impending energy issues and the range of climate change impacts that states are beginning to experience will dramatically curtail their previously claimed rights to pursue independent paths of energy production and greenhouse gas emissions that others may consider unsustainable or a contribution to climate instability (Bazilian et al., 2011). Nonetheless, interrelated and interdependent resource supplies and energy networks could be employed by states asserting political leadership as a basis from which to build complementary cleaner energy policies (Jacobsson and Johnson, 2000). All states will ultimately experience some energy resource security impacts as the effects of global climate change unfold and as such they all have an interest in supporting the efforts of others to transition to cleaner energy sources.

To some extent, precedents can be identified in earlier international climate change agreements, specifically those relating to the management and prevention of transboundary air and water pollution. Through these agreements, collections of states, typically neighbours or members of a region, identified collective interests and responsibilities for ensuring their access to clean air, rivers or oceans and their related resources. Such agreements have, at least informally, extended the scope of collective security by establishing shared responsibilities among member states for the well-being of their people and the quality of basic environmental resources they might utilise. While proximity provided common issues, identities and cultures as consensual foundation blocks this will undoubtedly prove more difficult to achieve at a global level.

Economic actors and an array of non-governmental organisations are interested in driving reforms towards sustainable energy production and consumption practices, and many are also willing to drive or influence political agendas and policy contexts (Alfsen et al., 2010). In the 21st century, energy markets are highly interdependent, as oil and other energy resources are bought and sold across the world. Consequently, most states are accustomed to a single energy market where high levels of reliance upon a limited range of resources create a highly charged political backdrop (Yergin, 2006). In the near future, these interdependencies and concentrations of energy production may exacerbate key security risks, but they are also likely to increase the effectiveness of international energy mitigation and adaptation approaches that are ultimately agreed upon (Anadon and Holdren, 2009).

States and non-state actors both have an interest in ensuring ongoing access to immediately useful energy supplies and long-term energy security. Underlying these motives is an appreciation that the certainty and security of energy supplies are the basis of international stability and therefore peaceful relationships between states. For many governments, energy security issues tend to revolve around efforts to limit the depth and extent of their political and economic vulnerabilities arising from dependence upon various fossil fuels for all aspects of production, including food, and maintaining large urban centres. While many states remain unsure about how to address these challenges, most readily agree that they increasingly need to reduce their 'risk of unmanageable disruption' through the climate-related impacts of carbon dioxide emissions 'from...fossil fuel burning' (Anadon and Holdren, 2009, p. 89). As energy use underpins every facet of desirable modern life in the provision and use of both goods and services it lies at the heart of efforts to mitigate the causes and effects of global climate change.

This broader international political context leads many to conclude that resource-related conflicts are likely in the 21st century (Peters, 2003). However, we tend to agree with those who believe that although political conflicts over resources (especially water) are almost certainly inevitable in coming decades, the international political community can do a great deal to diminish their scale and intensity (Wolf, 1998). Chief among these will be efforts to expand existing security and governance arrangements that increase the likelihood of resolving tensions and disputes before violent or armed conflicts occur. While this is likely to require collective political will on an unprecedented scale, it is worth bearing in mind that the 20th century produced sophisticated and effective collective security arrangements, including peace-keeping structures, to deal with similarly complex issues. We are hopeful that these might be further extended to support efforts to manage and alleviate conflict over resources. Ultimately, however, the success of such possibilities will rely upon the decisions and actions of coalitions of states and other political actors who pursue shared goals and who are prepared to modify their production goals and industrial practices. Although collective agreements are tremendously difficult to achieve, and are likely to be too slow in emerging to mediate all of the effects of climate change after so many delays, they nevertheless 'constitute the most effective way to ensure optimal emission reductions' (United Nations Committee for Development Policy, 2009, p. 23).

New energy pressures will increase insecurity among people and governments and challenge their abilities to provide political stability and economic prosperity (Anadon and Holdren, 2009; Yohe et al., 2004; Barnett, 2001). As climate change impacts, and policies to alleviate some of these, alter patterns of energy production and consumption, they will also change the nature of relationships between governments and their people and between states. These new political realities are as inevitable as increasing atmospheric temperatures and changed rainfalls patterns. It is likely, however, that some of the international security and peace management mechanisms that were created in response to experiences of international conflict in the 20th century might also be useful in reducing the risks of global conflict arising from energy shortages.

8
Water, Food and Fire

Introduction

Throughout the 20th century, higher atmospheric temperatures increased the intensity of the global water cycle, altering the processes of water evaporation, cloud formation, rainfall, river flow and groundwater storage, thereby changing 'its impacts upon regions and continents' (Huntington, 2006, p. 90). Further changes will continue into the future as atmospheric temperatures continue to rise. These changes in the global water cycle are altering the distribution of water resources which will, over time, also alter the suitability of many parts of the world for human habitation (Postel, 2000, p. 941). As is often observed, climate change has begun to make parts of the world wetter, more prone to flooding, with shorter periods of intense rainfall, while others are becoming more prone to severe and protracted periods of drought. Many scientists argue that even if urgent action is taken to reduce greenhouse gas emissions in order to slow the rate of increasing atmospheric and ocean temperatures, water redistributions will create major challenges for current patterns of human settlement, food production and industrial capacities (United Nations Development Programme, 2008; Parry et al., 2007; Pearce, 2007). If the build-up of atmospheric greenhouse gases is not reduced, and atmospheric temperatures increase more quickly, scientists predict that some 'areas of the globe...[will] become uninhabitable' (Diffenbaugh and Scherer, 2011, p. 616; see also Sherwood and Huber, 2010).

As awareness of global climate change expanded in the latter decades of the 20th century, scientists became increasingly concerned with the potential impacts of rising temperatures. Their observations led them to conclude that melting glaciers, ice and snow were likely to affect sea levels and ocean currents. It has only been relatively recently that changing monsoonal and other climatic processes that produce recognisable weather patterns, such as El Nino and La Nina, have also attracted widespread scientific interests (Collins, 2005; van Oldenborgh et al., 2005). These observed changes have generated an appreciation among the global community of the need to better understand the complexities of ocean currents and their cycles of heat and cold exchange. Ocean currents and wave patterns, global water and atmospheric temperatures, wind, storms and rainfall are interlinked and create global climates. Considering only the impacts of rising atmospheric temperatures misses many of the features and magnitude of 21st-century global climate change.

Water constitutes one of the most important life-sustaining natural resources and plays a variety of roles in the earth's abilities to regulate atmospheric and surface temperatures. It is as crucial as the presence of oxygen, and the stability of the solar system, to the earth's abilities to support the current diversity of plant and animal life. It is crucial to the survival of all species, enabling the growth of vegetation and the maintenance of human settlements. Without adequate rainfall, vegetation, fauna, rivers and lake systems die. Without adequate rainfall, societies struggle to produce sufficient food for their people, and grass, bush and wild fires occur more frequently, further threatening human settlements and species diversity.

Balancing demands for water presents a major political challenge for modern governments, and contests between them are likely to intensify in the near future (Serageldin, 2010; United Nations Development Programme, 2008; De Villiers, 2001). In the 20th century, differentiating between types of water, its potential uses, locations and availability became more commonplace and corresponded with increasing regulation and more intensive consumption. For instance, river water became differentiated according to its suitability and/or allocations supporting land-based irrigation, preservation of species, aquaculture industries, hydroelectricity production, suitability for consumption by urbanised populations and so on. Water became a subject of debates about pricing mechanisms and who might be held responsible for these, and who might have which kinds of access to water resources became international and domestic political issues.

The broader debates and concerns about greenhouse emissions and strategies for their reduction are also important for international responses to water redistributions and their associated economic, social and political challenges. Carbon trading schemes and other emissions reduction strategies will alter the dynamics and practices of global energy production and consumption, including food production and distribution. Carbon trading schemes might prove effective in reducing global emissions over the longer term but they do not provide an immediate or quick fix (see Chapters 6 and 7 for further discussion). It is already apparent that atmospheric temperature increases will have very largely resulted in the melting of polar pack ice and glaciers, and rainfall patterns will have changed, well before the benefits of carbon trading begin to change industrial practices (Dow and Downing, 2006; Houghton, 2004; Burroughs, 2003).

The global water cycle, incorporating oceans, lakes, groundwater systems, rivers, glaciers and polar ice, underpins climatic cycles, feeds seasonal rainfall, regulates surface temperatures and sustains life on earth (Jackson et al., 2001, pp. 1027–1030). Changes in the formation of rivers and clouds, increased ocean temperatures and altered currents within them, are affecting the manner in which water is located and distributed across the earth. Changed rainfall, river flows and lake system patterns are changing local and regional weather patterns, altering temperatures, humidity and moisture levels, and, in turn, dramatically altering the diversity, range, forms and distribution of vegetation and animal species across the globe (Figure 8.1).

- Floods
- Storm surges

- Changed river courses and volumes
- Changed river-water quality

- Loss of fertile soil
- Increased ocean salinity
- Challenges in water storage, drainage and flood management for urban communities
- Increased dust storms
- Increased floods
- Increased droughts
- Increased algal blooms in rivers and lakes
- Drying lakes
- Changing groundwater levels
- Changed distributions of waterborne diseases

Figure 8.1 How changes in water challenge human communities

Supporting human and other life into the 21st century will require political, economic and societal adjustments to these changes. Deciding how water will be used, whose rights to water will be privileged, and how the water market will be managed and regulated present crucial political decisions for all societies (Rogers and Leal, 2010; De Villiers, 2001; Postel, 2000). Governments will be confronted with difficult choices as they decide which kinds of water will be used for which purposes and seek ways of securing supplies. These will be important factors in determining who lives well, who experiences hunger, and what can be produced in which locations. Thus far, in matters of water regulation, governments have tended to favour human needs over the survival prospects of other species and environmental preservation. However, this could change as water redistributions require governments and societies to find new ways of producing sufficient food and secure water supplies for their people (Baron et al., 2002). Satisfying the water demands of the world's peoples, cities and production systems, while simultaneously protecting the ecological support functions of freshwater systems will be one of the most difficult and important challenges of the 21st century (Postel, 2000, p. 946).

Scientists anticipate major redistributions in global water resources in coming years on account of human population increases and changes to climate (Vörösmarty and Sahagian, 2000). 'Water scarcity has spread rapidly to many parts of the world... [and] the challenges water scarcity poses to food production, ecosystem health, and political and social stability will require new approaches to using and managing water' (Postel, 2000, p. 946). People, governments and intergovernmental organisations will become increasingly concerned to ensure sufficient food is produced to support growing global populations (Tubiello et al., 2007). They will also become increasingly concerned to ensure their societies are able to secure sufficient water supplies to support growing and increasingly urbanised populations (Rogers and Leal, 2010; De Villiers, 2001). Higher rates of urbanisation will mean the burden of food and biomass production falls to a smaller agricultural sector (Burroughs, 2003).

Threats to human communities

Current projections suggest many parts of the world will become vulnerable to recurrent floods, at least some of which will be particularly severe (United Nations Educational, Scientific and Cultural Organization, 2009; United Nations Development Programme, 2008). These pose direct impacts upon the lifestyles and locations within which human societies might reside into the future. Redistributions of water will also have impacts in terms of changing the availability and global distribution of land that is well suited to particular agricultural activities (Strzepek and Boehlert, 2010; Dow and Downing, 2007). Water's importance as a life-sustaining resource relates to specific questions of location, quality, flow, salinity and the relationship between rainfall and temperatures: when, where and how often rain falls makes an enormous difference to who eats which kinds and quantities of foods, and where communities can flourish.

Given the importance of water to all forms of human social and economic endeavour, it is reasonable to anticipate challenges to the political, economic and social systems within which people live. The political management, trade and patterns of exchange that have been created throughout recent history have relied upon expectations of water availability in particular locations. It is increasingly clear, however, that these distributions of useful water are changing. As the magnitude of global climate change impacts increase, so too will their effects upon water resources. It is anticipated that by 2030 available water is unlikely to increase by more than 10 per cent while the human population will increase by approximately a third (Jackson et al., 2001). Such eventualities will necessarily impede the forms of production that currently support the world's population.

In the past, many of the more catastrophic climate events, such as major land inundations from flooded rivers, tended to occur in rural and less populated areas. While catastrophic storms have occurred throughout recent history, many of their adverse and dangerous impacts upon human communities have tended to be experienced at comparatively low rates of frequency. This has been especially true among developed industrialised states who have largely believed that modern infrastructure and technologies made them impervious to the majority of extreme weather events. However, in the 21st century, global climate change impacts are challenging these beliefs as new patterns emerge. In Australia, across Europe and in other parts of the world, climate change–induced summer heatwaves, severe bushfires, and storm-related flooding will increase incidences of human deaths along with extensive losses of property, infrastructure and industries (Wang and McAllister, 2011).

Water distribution is changing because global climate change is altering the distribution of weather events and their patterns of occurrence (United Nations Development Programme, 2008; Dow and Downing,

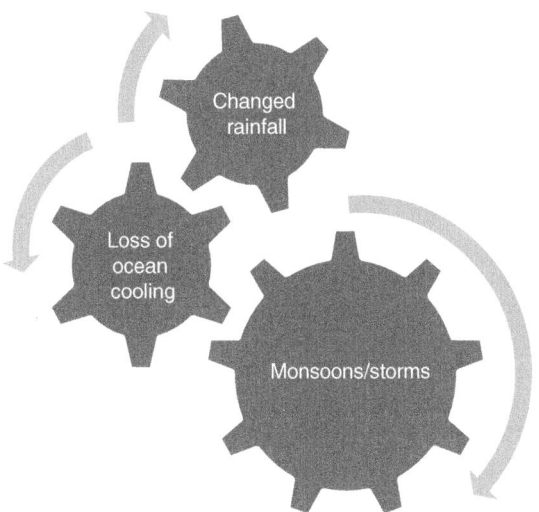

Figure 8.2 A changing rainfall system

2007; Houghton, 2004). Regional droughts and floods are increasing along with precipitation and evaporation that result in greater heat trapping (Huntington, 2006, pp. 90–91). These, and other climate changes, are beginning to affect the lives of almost all people but it is difficult to attribute specific storms, cyclones, mud or landslides, floods or droughts, as direct consequences of climate change (Huntington, 2006, p. 91). Although there is debate concerning the specifics of interconnections between the various processes and systems that determine rainfall, wind patterns, temperatures, ocean currents, river flow, ice and glacier formation, there is general recognition that such interconnections exist (Figure 8.2).

Drought, floods and crop failures which disrupt local communities are familiar experiences for many people in a wide range of geographic locations. Yet isolated severe weather events are rarely interpreted by distant authorities as indicating changes within the global climate systems. Nonetheless, it is possible to identify particular locations and peoples which have already encountered quite direct climate change impacts, especially among low lying island communities (Renaud et al., 2010; Bell, 2004; Suhrke, 1994). The threat of rising sea levels have forced government leaders and diplomats from low lying island states to begin lobbying larger neighbours to accommodate their people, presenting significant challenges for policy-makers (Renaud et al., 2010; Bell, 2004).

Coastal communities are not the only societies to be affected by changing water distributions and rising temperatures. Prolonged drought from the 1960s to the early 1990s in sub-Saharan African states in the Sahel region 'contributed to countless hunger-related deaths and unprecedented rates of migration from north to south, from rural to urban areas, and from landlocked to coastal countries' (Battisti and Naylor, 2009, p. 243; see also Kandji et al., 2006; Patz et al., 2000, p. 285). In 2003, severe heat events in Europe were directly attributed to reduced agricultural production and only global food trade, subsidies and insurance compensation maintained regional and global food security (Battisti and Naylor, 2009, p. 243). Nonetheless, 'the longer term challenge of avoiding a perpetual food crisis under conditions of global warming is far more serious .... It will be extremely difficult to balance food deficits in one part of the world with food surpluses in another, unless major adaptation investments are made soon' (Battisti and Naylor, 2009, pp. 240, 244).

The mass migrations of people following increases in ocean levels or prolonged drought are relatively easy to observe and track. However, myriad less immediately observable changes have occurred quite rapidly in recent years as the geographic distributions of many other species have altered. Human induced climate change and environmental changes are known to be impacting upon plant diversity and evolution with accelerated extinctions predicted over the next 50–100 years (Flannery, 2005; Tilman and Lehman, 2001). Many species have already become extinct as their abilities to adapt have not kept pace with the rapid rates of change in their habitat, food and water supplies.

Shifting and unpredictable water resources create stresses for human communities. Droughts and floods impact members of affected communities for many years beyond the immediate periods of inundation or water shortage. Populations socially dislocated and geographically displaced by droughts and floods may remain isolated from their former homes and communities for a decade or more. They experience a long-term loss of livelihood and an array of increased health risks (Patz et al., 2000). Droughts and floods destroy infrastructure, such as roads, dams, irrigation and sewage systems and, at times, hydroelectricity generators, as well as disrupting industries and communities. As observed by Patz, Engleberg and Last (2000, p. 283), '[e]xtreme weather events such as severe storms, floods, and drought have claimed millions of lives during the past 20 years and have adversely affected the lives of many more, causing billions of dollars in property damage'. As peoples, scientists and governments draw increasingly clear links between these events and climate change, governments will be expected to increase their efforts to protect their communities from the 'predictable surprises' of adverse weather events (Bazerman, 2006).

Seasonal temperatures, water supplies, soil fertility and healthy ecosystems all contribute to the well-being of human populations. In recent years, many more people have experienced the destruction of their homes, villages and cities as monsoonal paths, hurricanes and tidal surge patterns have altered (Overpeck and Cole, 2007; Patz et al., 2000). In addition, in many places bushfires are becoming more common and destructive features of long hot summers and threaten forests, grasslands, farms, towns and cities. As observed by Dube (2009, p. 169), interactions between 'fire, vegetation and soil is of particular concern because of the direct linkages to ecosystem goods and services and human well-being'. For many people in many places, previously occasional life-disrupting storms, floods, droughts and fires are occurring more frequently.

More people now live in increasingly fragile coastal and urbanised communities and 'even small amounts of sea-level rise would have substantial societal and economic impacts through coastal erosion, increased susceptibility to storm surges, groundwater contamination by salt intrusion, and other effects' (Alley et al., 2005, p. 456). More geographically dispersed people now risk severe water shortages and are likely to die of heat stresses, food and water shortages, floods and fires as global temperatures rise. Droughts and floods will disrupt established food and energy networks in coming decades, affecting the security and well-being of many people and the authority structures that provide social order. 'Food, water and energy resources are now at the stage where, by 2030, even quite small changes in climate could cause major famine' (Mason, 2003, p. 29). Evident throughout this range of issues is the increasing vulnerability of societies which have galvanised popular appeals for effective responses.

New political challenges

Storms, droughts, floods and the myriad health effects of changing weather and climate threaten the forms of societies that humans have constructed and thereby also challenge their visions of happiness, freedom and order (Patz et al., 2000). As noted by Diffenbaugh and Scherer (2011, p. 622), 'developed and developing economies are...likely to face unprecedented climate stresses even with the relatively moderate warming expected over the next half-century'. It is then worth bearing in mind the extent to which prosperity has enabled civil and human rights as relatively new political values across the world. At the very least, global climate change is eroding the range of choices and potential opportunities of many people. When coastal communities are inundated (Alley et al., 2005) or algal blooms kill productive fish stocks (Vitousek et al., 1997), they expose the limited abilities of governments to protect people. They also increase expectations that governments should be doing something more to mitigate and support climate change adaptation.

Predictability in weather patterns has long been important in enabling large-scale settlements which depend upon ongoing and reliable access to water and food supplies. In important respects, political, social and economic order among communities relies upon the abilities of producers, consumers and their networks for distribution and exchange to predict seasonal patterns. For instance, farmers across the world know that crop failures can occur at any point in the growing cycle. Irregular rainfall can cause the failure of seeds to germinate, stunted plant growth, or crop spoilage resulting in lowered yields. They also know that changing climatic conditions, such as extended periods of drought, alter soil fertility levels and affect long-term production even when rain falls.

Coastal communities

While changing rainfall and fresh water distribution and circulation patterns pose major challenges to many people in many locations, the very large proportion of people living within or near coastal communities face additional risks. Oceans are expanding their volumes as global atmospheric temperatures increase, causing both direct and indirect sources of increases in ocean temperatures. Melting ice, snow and glaciers are contributing to rising sea levels (Alley et al., 2005). Until recently, many scientists believed that melting polar ice had been largely restricted to the Arctic. Some political analysts suggested the potential benefits of new shipping channels and easier access to oil, gas, and mineral reserves, as well as immense reserves of fresh water, might offset the damage of sea level increases to the international political economy (Borgerson, 2008). However, in 2009, scientists became increasingly certain that substantial loss in pack ice had also occurred in Antarctica, evident in the unexpected and sudden shift of the east ice shelf (Scambos et al., 2003).

Rising sea levels have dramatic implications for the distribution of human communities and their political economies. As coastlines erode and higher levels of tidal surges occur, infrastructure, such as ports, airports, wind farms and housing, will be threatened or destroyed and agricultural land will also be lost. Rising 'sea levels will also lead to salt water contamination of groundwater supplies, threatening the quality and quantity of freshwater access to large percentages of the population' (Venton, 2007, p. 6). Tidal surges, coastal erosion, and salt contamination of freshwater sources will destroy some cities and demand huge financial and technological investment from others as they seek to retain their viability.

These impacts will affect incomprehensible numbers of people across many parts of the world. In some instances, densely populated and highly fertile lands will be destroyed and in some locations, entire societies may experience the loss of habitable land. A rise in sea level of 50 cm would seriously impact on the cities of the Nile Delta and could 'result in population displacement of about 2.0 million, loss of jobs of about 215 000 and value losses of over 30 billion' (El-Raey, 1997, p. 39). Similarly, large parts of island groups would be almost completely inundated in the event of a 1 metre sea level rise, especially in the Caribbean and the Pacific and Indian oceans. A 1 metre rise in sea levels 'would inundate low-lying areas, affecting 18.6 million people in China, 13.0 million Bangladesh, 3.5 million in Egypt, and 3.3 million in Indonesia' (Patz et al., 2000, p. 284). Conservative estimates of the effects of sea level increases suggest that more than '70 million people in Bangladesh, 6 million in Lower Egypt and 22 million in Viet Nam could be affected' (United Nations Development Programme, 2008, p. 18).

People are continuing to gravitate towards coastal areas, especially since these are often also 'areas of fast-growing economic development, but coastal erosion, rising sea levels, saltwater contamination, and potentially more powerful storms are expected to put these already threatened environments under increasing stress' (Dow and Downing, 2006, p. 64). Severe wind storms, 'can add 5 meters or more to the mean sea level' posing potential risks to many communities, including those that are not immediately located alongside coastlines (Dow and Downing, 2006, p. 62). Across the world, approximately '100 million people live less than one meter above mean sea level' and finding new homes for these will present major political challenges in coming decades (Dow and Downing, 2006, p. 24). Further, water redistribution impacts are expected to pose direct and recurrent threats for large numbers of very poor people in many different countries. It is predicted that '1 billion people currently living in urban slums on fragile hillsides

Table 8.1 Climate change impacts upon water resources

Ocean changes	Impacts	Consequences for human societies
Sea level rises	Coastal flooding Coastal erosion River contamination Delta inundation	Loss of habitable land Increased insurance costs Changed planning permits Loss of fertile land Loss of aquaculture industries
Changed heat and cooling exchange/current flows	Altered wind circulation Altered evaporation cycles Altered ocean ecosystems	Migratory species disruptions Species extinctions risks Changed storm paths Changed rainfall patterns
Warmer ocean temperatures	Lower rates of carbon absorption Fish and ocean species breeding cycle disruptions Altered ecosystems Acidification	Increased retention of atmospheric greenhouse gases Increased atmospheric temperatures Reduced viable fishing industries Reduced fish food stocks (tuna fishing inquiry) Loss of coral reefs

or flood prone river banks face acute vulnerabilities' (United Nations Development Programme, 2008, p. 18).

In 1990, it was estimated that approximately 200 million people, or 4 per cent of the world's population, lived on coastal flood plains subject to 1-in-1000-year flood events (Nicholls, 2003, p. 17; Nicholls and Mimura, 1998). Modest sea level rises were considered to increase the proportion of the global population affected to 6–10 per cent and increase the likelihood of flood events (Nicholls and Tol, 2006, p. 1081). Given the magnitude and extent of these global climate change impacts, it is especially important to acknowledge that 'around 40 per cent of the world's population lives less than 60 miles from the coast', placing them within easy 'reach of severe coastal storms' (Dow and Downing, 2006, p. 24) (Table 8.1).

Urban water upheavals

In recent years, urbanised populations have experienced first-hand some of the effects of altered rainfall patterns and their flow-on effects. Hotter, drier summers have taken their toll on people across the world and exposed their vulnerability to heat and water shortages as evident in the loss of life across Europe and North America in the heatwaves of 2003 and 2006. Despite regional variations, contributing factors in heat-related mortality rates are known to be urbanisation, population density, urban GDP and proportions of population aged over 65. At a global scale the risks of heat-related deaths rise by between 1 and 3 per cent per 1°C increase in high temperatures (Hajat and Kosatky, 2010). Extreme heat events are known to have acute impacts on human health, agricultural production and food security (Diffenbaugh and Scherer, 2011, p. 615).

Developed states face additional political and social challenges in relation to their current water consumption practices to support and enable recreational and leisure activities. Some of the water used in these forms of recreation, entertainment and leisure is consumed in direct forms, such as filling swimming pools and watering sporting fields, while much is also consumed in fuelling the electricity used in lighting, heating and cooling these facilities. While the volumes of water utilised by developed states in recreational activities are less than the volumes deployed for energy and industrial production, they nevertheless add significantly to per capita consumption rates (Mekonnen and Hoekstra, 2011). Curtailing this use of water then has direct impacts upon lifestyle practices and qualities of life as gardens, parks, swimming pools and sports fields increasingly become a luxury and costs restrict social access and availability.

Conflict over water consumption within communities is likely to trigger significant electoral backlash as well as civil and social conflict (De Villiers, 2001). In coming years, more people will experience more restrictive water consumption policies and increasing costs as water prices are amended to reflect relative scarcity (Mekonnen and Hoekstra, 2011). These unevenly distributed experiences are likely to cause added tensions between and within communities (Kasperson and Kasperson, 2001). In these and other ways, global climate change and its impacts upon water distributions presents popularly elected governments with unpleasant political choices that demand new approaches to framing lifestyle aspirations. Policies restricting water consumption might well be responsible, but unless these are evenly experienced and supported by educational measures and investment in new infrastructure, they are likely to cause political disenchantment. This is especially true in localities where shortage, hardship and sacrifice are perceived as unevenly experienced due to government policy or pricing mechanisms that are considered to privilege only some members of a society.

Managing unpredictable risks

Global climate change is increasing the range and frequency of fire risks as the incidence of drought increases across both tropical and subtropical climate zones (van der Werf et al., 2008, p. GB3028; Hoffmann et al., 2002, p. 2052). In these regions, established forests and grasslands have provided valuable carbon sinks and earth cooling zones. However, climate forced changes in rainfall patterns and diminished ground water levels dramatically increase the likelihood of fires with implications for human land-use interactions (Dube, 2009, p. 163). Climate changes also influence the frequency and relative heat intensity of fires (Granström and Niklasson, 2008, pp. 2353–2358; Power et al., 2008, pp. 887–907; Carcaillet et al., 2007, pp. 465–477; Pausas, 2004, pp. 337–350), thereby also increasing the difficulties of extinguishing them and/or preventing their spread. In combination, climate changes and human activity have significantly elevated the risks of fires to people and industries in many places (Archibald et al., 2009, pp. 613–630; Syphard et al., 2007, pp. 1388–1402).

One of the challenges faced by states and the international community in responding to changing water supplies, increased fire risks and new challenges in food production, concerns the persistent limitations of specific knowledge (Dube, 2009, pp. 161, 165). The complex interactions between fire, vegetation and soil are influenced by both climate and land use (Van Langervelde et al., 2003, pp. 337–350; Siegart et al., 2001, pp. 437–440). It is important to understand these interactions now because '[f]ire is a subject of global significance' and there is rather limited knowledge of 'how it will interact with climate change' and how these will create feedback systems (Dube, 2009, p. 161; see also Okin et al., 2009, pp. 237–244; van der Werf et al., 2008, p. GB3028). It is known that fire patterns are influenced by both climate changes and human interactions with the environment. Climate influences conditions for fire while human land use and activity influences fire frequency, scale of burnt areas and fire intensity (Grissino-Mayer et al., 2004, pp. 1708–1724; Veblen et al., 2000, pp. 1178–1195). Similarly, insurance companies have begun to reinforce public concerns by imposing substantial premium increases upon businesses and residences in fire risk areas. As such views and perceptions find public expression, governments may well find that there are many across diverse electoral groups whose patience for slow, incremental, and largely ineffective, targets has run out (ABC Radio News, Melbourne, 19 November 2009).

Substantive political problems persist in the debates and attempts to understand, define and respond to the implications of changing water distributions, rainfall and related climatic patterns. The political problems of ensuring that sufficient food is produced and distributed and societies are not overwhelmed by problems of fire and water cannot be achieved by single sets of targets because the problems themselves are too complex and too varied across geographic zones and seasons. They also comprise interrelated events and processes which increase the difficulties of identifying universally appropriate management, adaptation or mitigation strategies. These are problematic issues for international politics where actors are better equipped to deal with general principles. As with other aspects of global climate change, altered water distributions raise new questions and expectations about what governments can or should do to secure the well-being of their peoples.

Seasonal uncertainties

Overall, across many parts of the world dry seasons are becoming drier, and wet seasons are becoming wetter (Garnaut, 2011; Venton, 2007). Also, in many locations, wet seasons are becoming shorter while dry seasons are becoming longer. Shorter, wetter winters might sound something of a boon to those who dread long dreary winters in temperate climatic zones. However, shorter wetter wet seasons in such regions dramatically increase the risk of severe fires during dry season, due to the accumulation of dry and combustible fuels (Evett et al., 2007, pp. 318–330). They also increase the risks of floods occurring in wet seasons as localised heavy rainfalls increase erosion, change water courses and breach drainage infrastructure (Garnaut, 2011).

The 'global intensification of the hydrological cycle' will result in many currently inhabited regions of the world facing 'heightened risks of more extreme and frequent floods and drought' (Venton, 2007). Climate change effects are dispersed such that some 'areas will experience less annual rainfall; in others it will be less predictable, with seasonal rains failing to materialize, or arriving with such ferocity that they create dangerous floods' (Dow and Downing, 2006, p. 56). For a great many people, unpredictable rainfall and access to water supplies will become major disruptions to the industries that support their lifestyles and the governments that maintain social order. At both ends of the wetter and drier spectrums, more people will suffer, and many will experience both wetter and drier conditions across changed seasonal cycles with effects upon human health. Recent experiences of social and health impacts from heatwaves, floods and droughts, provide insights into these. For instance, 'heat causes hyperthermia, cold causes hypothermia, and droughts cause famine. Injuries, displacement, and death result from floods, hurricanes, tornadoes, and forest fires...climate and weather affect the distribution and risk of many...infectious diseases', including those that are borne by water, food and wind (Frumkin et al., 2008, p. 435).

Across the world, but especially in those regions that are already dry, problems of drought and desertification will intensify. Higher atmospheric temperatures will increase the incidence of drought, and increase the numbers of people lacking access to adequate water supplies. These problems are expected to be most intensely experienced in southern Africa, North Africa, Central America and central Asia (Venton, 2007). Long-term water shortages are likely to be experienced in some previously water secure locations, such as Israel and south-east England where importing water has been 'suggested' as a mitigation strategy (Dow and Downing, 2006, p. 56). Further disruptions to communities around the world are likely to arise from increasingly rapid melting of ice, snow and glaciers as global atmospheric temperatures continue to rise.

In major rivers, glacial, snow and ice pack melting will increase 'river flows in the spring, but may reduce summer flow', including in the Indus and Ganges river basins where erosion and water contamination are also expected to affect the well-being of those who depend upon these rivers (Dow and Downing, 2006, p. 56). In these, and other rivers, glacial melting 'will increase flood risks during the rainy season, and strongly reduce dry-season water supplies to one-sixth of the world's population' (Venton, 2007, p. 6). The magnitude of such changes defies imagination and exceeds the capacities of states to manage in an orderly manner. For instance, the food security of a great many people relies upon the Indus and Ganges rivers whose flows are expected to be compromised by climate change (International Union for Conservation of Nature, 2011). These realities are exacerbated by the complex processes that underpin the distribution of freshwater resources which create 'a different water regime' for each continent, where each imposes particular constraints upon the 'population that continent can support' (Burroughs, 2003, p. 152). Coming to terms with the fact that 'precipitation alone will not meet the needs of a growing population over the next 50 years' will require governments and others to deal with disruptions to the lives of many people as water shortages, changed river flows, and changed soil fertility patterns have political and social ramifications that flow far beyond state borders and regional associations (Rogers and Leal, 2010; Burroughs, 2003, p. 152).

Fresh water resources and disease

Widespread changes in the distribution of rainfall and water resources are expected to compromise the quality of freshwater supplies in a significant number of locations. Where lower rainfall reduces water levels and river flows, increases in the relative levels of contamination from sediment, algae and bacteria are likely. Lower summer seasonal flows in rivers affected by spring glacial melt will also increase their levels of contamination. Specifically, heavier seasonal rainfalls will increase the 'risks of waterborne diseases such as cholera, typhoid and dysentery, and mosquito-borne diseases including malaria and yellow fever' even when these rains do not result in major flooding (Dow and Downing, 2006, p. 60). These health risks might be anticipated to severely impact human populations that depend upon river-based water sources.

Other less obvious consequences of floods arise from the soil-borne diseases that are transported and exposed by flood waters. These include 'anthrax, and toxic contaminants such as heavy metals and organic chemicals' (Dow and Downing, 2006, p. 61). These negative impacts of global climate change are unlikely to be limited to small geographic regions or to be easily addressed by adaptation or mitigation strategies. Within the foreseeable future, many of the world's most vulnerable peoples will lose at least a proportion of their already limited water supplies. Their abilities to adapt to these changes and to find ways of making the regions they currently occupy suited to their ongoing habitation will be limited.

One of the tremendously confronting challenges concerning water resources is the basic reality that water cannot be manufactured. While new technologies can improve water use efficiencies they cannot overcome the consequences of drought, or prevent the spread of deserts, or the melting of glaciers. It is now readily apparent that global climate change in the 21st century will alter the availability and quality of water across the world. What is less clear is how governments, producers, industrial sectors, economic actors of various kinds and human societies in various locations will seek to mitigate these consequences and adapt to them.

Reducing greenhouse emissions will be an important component of international climate change mitigation efforts, but even if these can be universally and readily agreed upon, changes in water supply distribution, quality and rainfall patterns will continue throughout the 21st century. Adapting to these water challenges will require the mass relocation of human populations and the redistribution of agricultural land. Orderly and timely responses to these changes will require collective political will. New forms of political authority and sources of collective responsibility will be important components of efforts to manage the effects of water redistributions. These issues present enormous logistical challenges, even if international common interests can be identified and policies implemented in pursuing internationally agreed goals.

Food, water and adaptation

The political and economic implications of the changing global distribution of rainfall, fresh water supplies, rivers and coastlines range far beyond the problems of creating a new global food supply, although this too presents complex sets of practical and political challenges. While some of these challenges can be resolved by establishing new sites for water catchment and electricity generation, such initiatives require enormous financial investment at a time when many other demands are placed upon public expenditure and infrastructure development (Rogers and Leal, 2010; Burroughs, 2003). The infrastructure, established to service the well-being of human societies, under one set of climate conditions is likely to impose constraints on how we imagine dealing with a new set of climate conditions. Changing rainfall patterns will alter states' abilities to meet their future energy needs as rivers change course and flow rates. These changes raise doubts about the long-term viability of hydroelectricity generation in some instances, and render water storages and other infrastructure doubtful contributors to the well-being of the societies they were intended to support (Rogers and Leal, 2010).

The effects of changing rainfall and river flows will not be evenly experienced among states. Especially in the short term, some states are anticipated to benefit from warmer conditions and longer growing cycles. However, many of these states are not major food producers, so their contributions towards a global food supply will be modest. In addition, redistributions in rainfall and water supplies may have substantial negative impacts upon global food supply (Battisti and Naylor, 2009). Even more disturbing, changed rainfall and monsoon patterns across Asia are likely to dwarf these impacts, creating human tragedies through both direct and indirect destruction of human communities. Floods, droughts, famines and outbreaks of diseases that affect the lives of increasingly greater numbers of people annually are among the forecast effects of changing weather on public health (Patz et al., 2000).

The social, economic and political impacts of floods, droughts and severe storms often last well beyond the immediate efforts entailed in cleaning up cities, towns or villages after storms, or rebuilding destroyed housing and industries. Flood waters can take months to recede, and even when they do, dislocated or displaced people may be unable to return to their former towns and homes. They create long-term challenges for all facets of human societies and political systems. Reconstructing cities and towns can require years of sustained effort and investment following storms or flooding. Rebuilding communities after storm or flood damage frequently necessitate government initiatives to support the widespread replacement of damaged infrastructure, housing, commercial centres and social services, as well as the management of dislocated populations.

Severe floods and droughts present additional costs to societies in terms of their disruption to education and other social programmes. For instance, children are 'more likely to be malnourished if they [are] born during a drought' (United Nations Development Programme, 2008, p. 16). Similarly, floods can have lasting impacts upon the lives of those whose childhoods are disrupted, including abruptly ending their school attendance, resulting in poorly educated local populations with diminished life opportunities (United Nations Development Programme, 2008, p. 17). These impacts are not restricted to developing states, although developing states tend to experience more frequent climate disasters and lack the resources to make rapid recoveries (United Nations Development Programme, 2008, p. 16).

Knowledge and political will

As scientific knowledge and understanding of changes in the global hydrological cycle expands, the international community is struggling to come to terms with the range of challenges that such changes might create for the viability of contemporary societies. Their task is enormous and this would be so even without the additional complexities that might arise from rapidly changing rates of water resource exploitation currently being pursued in various locations. In summary, many human communities have recently begun to interact in increasingly exploitative ways with the rivers and oceans that support them. Increased intensity of ocean fishing and aquaculture industries have disrupted breeding cycles of certain species and/or subjected certain fish populations to levels of overfishing that have resulted in extinction or dramatic depletion of their numbers (International Union for Conservation of Nature, 2011; Burroughs, 2003, p. 42). However, even greater disruptions are in train as various states seek to support their increasing urbanised populations by expanding their available water supplies through the construction and operation of desalination plants. Such plants might produce water that is 'suitable for most urban purposes' at prices that are broadly 'comparable to importing freshwater', but they also rely upon the use of fossil fuels and contribute to greenhouse gas emissions and thereby further disrupt global hydrological cycles (Burroughs, 2003, p. 189).

Conclusion

Changes in water distribution will alter the geographic range and location of forests, producing further shifts in rainfall patterns, groundwater storage levels and run off into streams, rivers and water storage facilities (Rogers and Leal, 2010; Dow and Downing, 2007). Further changes will occur in the hydrological cycle arising from ongoing vegetation loss due to continuing urbanisation, drought, rising soil salinity levels, increasing atmospheric temperatures and changing season cycles across the world. Redistributions in the global water supply will be especially disruptive to human communities which have developed around sources of dependable and predictable supply. Climatic changes that make water either less dependable or predictable will impact upon the sustainability of cities, agriculture, industry and particular forms of lifestyle. Floods and droughts are widely recognised to have short- and long-term impacts upon societies, with implications that spill across generations (Smith and Petley, 2009; Patz et al., 2000). As they become more frequent, severe or prolonged occurrences they will generate increasing remediation costs that will have political, social and economic consequences.

Water constitutes one of the staples of life and changes to its availability produce very different futures for the natural world and human communities. Whether or not climate change science is accepted as an indicator of human induced climate change or merely that human activity is contributing to climate change is largely irrelevant. Natural cycles are redistributing water across the globe and this will have significant implications for the long-term future of life on earth. An overabundance or a shortage of water necessarily changes the fundamental biological processes that support life and as a consequence both the natural world and human communities will need to evolve and adapt to their new environments.

Responses to changes in the distribution of water by states and the international community will likely generate further political consequences. Many of these might not be contained within the borders of individual states, and states might prove poor structures for maintaining social order as water redistributions create new political, economic and social challenges. The processes and systemic features of climate change, where rain falls, and how strongly wind blows, do not recognise political boundaries or the locations of communities. Borders were created to divide resources and sources of authority between human polities. They remain blind to the characteristics and systems of nature (Mische, 1989).

Within states, natural and historical distributions of water often account for the historically established demographic dispersal of peoples, industry, agriculture and settlements. New seasonal variations that bring too much water, or not enough water, to particular locations result in hardships that have to be addressed through combinations of new infrastructure, shifts in population densities or reallocations of water. Each of these responses has political aspects that may not respect democratic processes. Even when they do, population densities are likely to produce outcomes that are not equitable. For instance, decisions to divert river and dam waters to cities, in order to support expanding urban populations, occur at the expense of less densely populated rural areas that require water for agricultural purposes. Such decisions have both short- and long-term consequences for ecosystems, future climatic patterns and the composition of societies. The tensions arising from responses to climate change and water redistributions among and between societies are likely to impact upon the abilities of states and the international community to maintain social and political order within and between states.

An overabundance of water through increased rainfall, flood, storm surges, rising sea levels or severe storm events create a host of health problems for human communities. A lack of water giving rise to desertification, reduced agricultural output, dust, soil erosion and species loss, likewise have health impacts. Perhaps the most dramatic of these are increased incidences of fire which have the potential to destroy forests, grasslands, infrastructure, communities and lives. As existing habitats become drier and more prone to fire, they present further challenges to the distribution of human populations, and the ways in which these are provided security. Additional challenges are likely to emerge in terms of the expectations placed upon states in providing security for their people, territories and resources. Fires produce atmospheric pollution, destroy natural carbon sinks and change natural environments which all impact upon natural processes that mediate climate changes. In this sense, fire is a by-product of redistributions in global water availability that both contribute to, and are a consequence of, climate change. In both cases, it is an increasingly common and severe aspect of climate change that has significant implications for existing communities and their future development.

9
Solutions, Ideas and Institutions

Introduction

Ideas about transforming human societies contributed to the industrial revolution and entrenched beliefs in progress and civilisation alongside the development and adoption of new production practices. These developments were supported by additional ideas concerning rights and prosperity across the world. Modern states were created through these ideas and early states were supported by additional ideas concerning legitimate authority and capacities to exert an independent rule of law over territory (Philpott, 1997). Through these developments, states became key political entities as they were attributed particular status and rules were established for their conduct.

For several centuries, these ideas have provided states with the language of engaging with others and enabled them to claim rights to engage in conflict or cooperation. They have also shaped states' collective and individual responses to the complex challenges posed by global climate change. As discussed in Chapters 2 and 7, these ideas have also indirectly created many of these problems. They established expectations of prosperous economic conditions that led to the range of human activities now identified as having contributed to global climate change and its associated ecosystems depletion.

The absence of unified, politically orchestrated responses to global climate change makes it appear that individual states are looking for scapegoats to avoid taking responsibility. While this may well be the case for some states, it is more likely that many political leaders are struggling to understand how to respond to these problems. Many of the delays in progressing a comprehensive response to climate change indicate that political leaders are uncertain of their knowledge of the problems, or potential solutions to them, or both. Climate change means we no longer know how to envisage our future because we do not know what we might expect in continued production or what future progress might entail. At present, we, and our governments, lack new ideas for transforming human societies and this leads us to wait for greater certainty before acting.

Climate change requires fundamental ideological responses. These are ideas associated with what scholars call a new 'creedal period' that might set new international expectations of authority and establish new norms of orderly conduct (Tannenwald and Wohlforth, 2005, p. 4). The history of ideas suggests the international political community and its constituents will ultimately produce the means for responding to global climate change. However, these ideas must carry a grander vision than focusing on ways to improve the efficacy of solar energy storage, or assessing the merits of wind power, or debating whether covered irrigation channels will be sufficient to overcome evaporation to meet ongoing water needs.

Problematic ideas and interests

The key ideas of modern industrialisation have been important 'causal factors' in contributing to climate change and they continue to make it difficult to generate comprehensive responses. One of these difficulties concerns the need to develop effective substitutes for the dominance of policies based upon expectations of continuing economic expansion, including increased global production based on fossil fuel energy sources. This is especially problematic for states whose nature and existence is predicated upon ideas of national interests that rely upon maintaining high growth lifestyles. International environmental policymaking remains stalled within the constraints created by states seeking to maintain and fulfil earlier aspirational promises to their citizens and securing their place in a competitive political economy.

As the world grapples with the potential consequences of climate change, ideas concerning the nature of political and social organisation, institutions and society are thrown into a state of flux. The political dimensions of climate change problems are unprecedented. This is the first time in history that all peoples, all forms of societies, in all quarters of the world are facing similar complex physical and material challenges arising from a set of dominant, progress fixated, ideas. Throughout the 20th century, as newly independent states sought to achieve industrial progress, they largely emulated the successful strategies of western industrial development and production (Mansbach and Rafferty, 2008, pp. 570–592). Although various development and modernisation efforts failed, international responses (as well as the hopes of many developing states) have remained premised upon unlocking the benefits of industrial development (Kukreja, 2005; Sachs, 2005).

Current expectations of climate change consequences necessarily set this entire project aside. In the face of the consequences of climate change, the question for developing and developed states alike should become, 'how can we best protect ourselves from the worst of these?' International responses to this question are divided by debates about the accuracy of specific predictions, the particular responsibilities of individual actors, and whether new ideas and technologies might fall short of solving the problems of global climate change, but enable ongoing capital accumulation and production.

Knowledge, technologies and innovation are central to the identification, agreement and pursuit of solutions to the problems posed by climate change and shortages in water and energy resources. Articulating and applying ideas to shape international responses to emerging climate change relies upon the abilities of governments, international actors and individuals to cultivate shared visions. Such an outcome is likely to arise, in part, from explorations of the processes and implications of restructuring societies and their authority structures. Knowledge and innovation then become important means for exerting influence among states as the international political community seeks new opportunities for preserving orderly relations (Young, 2002). Consequently, new ideas might become sites of power and influence within the international political economy (Haas et al., 1993).

Global issues such as climate change, water redistributions and challenges to orderly relations between states apply considerable pressures to the traditional principle and practice of sovereignty. Modern governments, shaped by traditional views of sovereign authority, perceive their interests as conforming to oversimplified dichotomies such as the 'internal' and 'external' realms of states (Doty, 1996). However, sovereignty is an evolving concept, and climate change shows how the autonomy of states cannot now be conceived of as an inflexible right. Indeed, sovereignty can be a problematic principle: it has contributed to conflict over territory, ambiguous criteria for statehood and divided social identities. Its adoption has produced considerable diversity in the nature of authority and influence that states seek to exert over each other and contributed to contemporary security risks (Weiss, 2011; Walker, 1991; Mische, 1989).

The role of ideas

Ideas underpin and inform decision-making (Nyberg, 1993, pp. 29–41). They shape the choices made by people, states, political associations, economic actors and institutions. Widely accepted ideas enable us to make sense of our observations, plan our activities and identify the causes of events (Shapin, 1994, pp. 3–41). They enable us to understand and explain the nature of the world in which we live and to make decisions concerning how we form and maintain societies, and interact with our environment. They shape how we conduct various production processes and create sovereign authorities that we believe will protect us from threat.

Widely accepted ideas are important sources of order in the contemporary world. They form the basis of governmental authority and underpin the property and representational rights that many people enjoy. Over time, ideas are subsumed into accepted modes of conduct, enabling conventions, customary practices, rules, law and political structures to be established. New phenomena, such as the challenges emerging from global climate change, heighten our awareness of the importance of ideas in enabling social and political order.

Shared ideas provide the basis of group activities, including the formation and maintenance of societies where they might attract unanimous support or remain bitterly contested (Shapin, 1994, pp. 3–41). This is especially the case in politics as it is from ideas that passionate consensus or discord arises to motivate responses. Contested ideas about climate change conform to this model. Appreciation of the consequences of global climate change and the likely effectiveness of responses have energised actors both domestically and internationally to re-examine the real world role of ideas (Raustiala, 2001, p. 104). The extent to which ideas about global climate change become commonly held then determines the abilities of key political actors to align their interests and responses.

After several decades of accumulating and refining knowledge about climate change consequences, and occasionally seeking collective strategies to alleviate and manage them, states seem to have achieved little progress in overcoming fundamental political dilemmas: Who can be held responsible? Which states should bear the costs of these consequences? Who should bear the costs of beginning the large-scale social and economic changes required to mediate the consequences of climate change? Will the authority of states survive? The growing preoccupations of many states in seeking to address or deflect these questions reveal the scope of climate change impacts.

Ideas of responsibilities and rights are important alongside ideas of cost and burden sharing for states debating possible responses to climate change and water redistributions. Ideas of progress and preservation must be reconciled with ideas of managing scarcity and rapid social change. In these areas of human social and political organisation, ideas provide instruments for making sense of observations, values and visions. Ideas enable observers to assess and develop new interests on the basis of their responses to knowledge. In this way, ideas create order and enable understanding of new phenomena, new knowledge and new problems.

In terms of addressing the effects of global climate change, an important role is played by the transmission of ideas, especially in relation to their common acceptance through which the national interests of states might be brought into closer alignment. Many commentators see this as a necessary precondition for creating the levels of international cooperation required to achieve effective mitigation and adaptation strategies. Among them are those who 'hope... that the international community will... be able to create the necessary institutions and agreements to restrain' states and other actors from pursuing interests that damage 'global well-being' (Luterbacher and Sprinz, 2001, p. 12). As global climate change impacts have become visible, it is important to examine the roles played by different types of non-state actors in addition to states.

Climate change is affecting all aspects of the world's weather patterns and altering the distribution of water and challenging the present forms of domestic societies (United Nations Educational, Scientific and Cultural Organization, 2009). These changes are producing urgent imperatives for new policies, structures, norms and rules for the behaviour of all international actors. In responding to climate change, states cannot act independently because their decisions and actions lack meaning without broader support. It is increasingly accepted that states cannot unilaterally control their physical environments nor independently protect themselves (Edmondson, 2011; United Nations Development Programme, 2008). Climate policies influence 'a host of other environmental and social problems' because they are 'intimately linked with energy, transport, and forestry policies' (Raustiala, 2001, pp. 98–99). Responding to climate change impacts will change 'consumption and production patterns throughout the industrialized world – and to a lesser extent in agricultural societies' (Raustiala, 2001, pp. 98–99).

The political challenge

Effective responses to the threats posed by climate change depend upon understanding its possible impacts and the views held by people and states concerning their responsibilities for protecting past achievements and future opportunities. Climate change now pervades both international and domestic political and social debates about the future, and has galvanised the imaginations and activities of a broad host of actors. It demands immediate attention and action. Its impacts upon international and state development, prosperity, security and order are manifold even though some consequences remain clouded in uncertainty.

The principal political concerns of the 20th century may be looked back upon as a contest between ideologies to determine how best to organise human societies and the utility of war in settling disputes between states (Fukuyama, 1992). These concerns will continue to be important for the international political community as states, their political and economic institutions and people respond to the critical issue of the 21st century. However, they will also stimulate new contests for political economic and ideological dominance. Resolving these conflicts will be important to climate change mitigation and adaptation which will otherwise risk repeated derailing by short-term marginal political issues. Nonetheless, our abilities to address these concerns can now be premised upon meeting the challenges of climate change.

Global climate change has been recognised as an international challenge since the middle of the 20th century but this has often been limited to small and specialised communities of scholars, scientists, activists, international institutions and government technocrats (Boehmer-Christiansen, 1996; Vogler, 1996; Young, 1994). In more recent decades, however, it has become an issue of broader concern. A partial explanation for this lies in the politics of the late 20th century. For much of that time, political attention and energies were expended upon a range of security and economic development concerns. For many developed states, Cold War manoeuvres and nuclear war threatened the security of their achievements even as their levels of industry-led affluence increased (Leaver and Richardson, 1993). For newly independent states economic imperatives and issues arising from the need to establish their own political identities were, and for many remain, all consuming.

Solutions, Ideas and Institutions 161

It has only been since the end of the Cold War, and the era of liberal triumphalism of the early 1990s, that many in the developed world shifted their attention to consider the next global challenge to international order (Fukuyama, 1992). The events of 11 September 2001 supported suggestions that a purported clash of civilisations might pose the next central threat to the international political community (Huntington, 1993). Since then, however, the war on terrorism has emerged, not as a global challenge, but as a rather dismal and degrading conflict between a self-selected number of states. This is not to deny the possible threats arising from international terrorism, or the implicit dangers of war. Rather, it reflects recognition that such issues are not truly global and do not impact upon or hold importance for all people in all states.

A further explanation for climate change becoming recognised as the global threat of the 21st century lies with the diverse range of actors who have become involved in its examination and reporting. (See Chapter 4 for further elaboration of the roles of new actors.) Especially among democratic and industrialised states, it has become a household concern as either a genuine threat to the future, or as a threat to standards of living and lifestyles. Passionate and energetic debates occur within daily media across states and continents indicating the extent to which climate change has become part of the context of contemporary life and political organisation. Ideas rather than facts underpin these developments since they provide the means by which people, their governments and other sources of organisation seek to assure themselves of security, longevity and well-being.

Responding to climate change impacts requires political innovations, insight and resolve as well as scientific expertise and technological advances. The complexities and uncertainties they express ensure that 'scientific debate will continue to be a key part of the political process' (Rowlands, 2001, p. 62). Climate change converts and sceptics alike are able to write endlessly either for or against the extent of the climate change threat, the range of potential responses, and the significance of human activities as contributing factors. As a result, new forms of authority and structures to ensure policy implementation will be as crucial as alternative energy sources and initiatives to limit greenhouse emissions.

Although the issues associated with climate change and its effects are primarily geophysical problems, their impacts and consequences play out in international relations by changing material circumstances and the conditions within which societies can be maintained. Sovereign states comprise the fundamental form of social and political organisation and they relate to each other through established patterns and institutions associated with recognition, legal equality, trade and diplomacy (Edmondson and Levy, 2008; Keohane and Nye, 2001). The abilities of states to exercise authority over people and territory sit alongside their needs for economic resources, markets and physical security. States negotiate, bargain, conduct diplomacy and otherwise engage with each other in order to maintain their status within the international political community and to ensure the continuation of their domestic societies (Edmondson and Levy, 2008).

Ideas and resilience

As socially constructed entities, sovereign states evolve in response to the concerns, obligations and duties that emerge through their interactions (Biersteker and Weber, 1996). The interests of domestic and international political actors and their assessments of impending security threats drive their actions (Philpott, 1997; Finnemore, 1996). States rely upon ideas and express their ambitions by interacting with numerous actors. They endeavour to represent and address the concerns of their domestic constituencies, giving rise to observations that states inevitably pursue self-proclaimed national interests. While states sometimes claim that they are exerting their will in international relations through these interactions, the reality is that they also remain responsive to the political pressures exerted by others (Edmondson and Levy, 2008).

States create new shared interests and values through international relationships with others, including non-state actors. Their preferences, like those of other international actors, arise from the interplay of ideas located both within and beyond their borders. These observations have led to understandings of states as subject to changing interests, structures and relationships that give rise to 'new preferences' (Finnemore, 1996, p. 10). As a potential trigger to major restructuring and political disorder in the international political community, climate change emphasises the importance of recognising that states' interests are malleable and receptive to instruction from external sources (Edmondson and Levy, 2008; Finnemore, 1996, p. 11).

The international political community is not a full society although it displays some societal features. It lacks an overarching authority and displays only some of the features normally associated with societies, especially in relation to loyalty towards others and institutionalised recognition of interdependencies (Edmondson and Levy, 2008; Luard,

1990; Kratochwil, 1989). An array of formal and informal rules govern the relationships between states, and prevent unrestricted violence, instability in international agreements and insecurity arising from the independent authority of their respective governments (Bull, 1995, p. 64; Lipson, 1991). This lack of a common government means that the capacities of the international political community reside in 'common norms, rules and identities among states' (Buzan, 1993, p. 339). These patterns of behaviour and the ideas that underpin them support norms (which amount to accepted forms of behaviour and political identity) and enable institutions to be created by states to guide their future behaviour and actions.

Ideas about order

Norms and institutions exist to support orderly conduct. They create predictability, expectations and standards of appropriate behaviour that inform states' abilities to articulate national interests, form alliances and participate in collective decision-making activities. Some of these international principles are older than many of the political ideas and values of the contemporary world. Some were developed to ensure that agreements between states should be kept to ensure the safety and appropriate conduct of diplomatic envoys (Williams, 1989). Others were developed to ensure mutual goodwill and trust in maintaining alliances to support orderly conduct among states (Edmondson and Levy, 2008, pp. 182–202).

These kinds of principles create expectations of behaviour and thereby reduce the insecurity that states experience. International institutions enable states to share information, to set terms of acceptable forms of behaviour and provide guidance regarding legitimate actions (Brennan and Buchanan, 1985, p. 10). Predictability and expectations create standards of behaviour. In turn, states work to maintain these standards in order to ensure predictability and stability exist as a feature of international society.

This standardisation is important because it facilitates co-operation among states, enabling them to judge the behaviour of others in terms of broader patterns of acceptability (Nadelmann, 1990, p. 481). Standardised expectations also provide states with advance notice of the range of actions that others might imagine acceptable, enabling them to condemn 'those that fall below the standard' (Luard, 1990, p. 202). International institutions and their accepted forms of behaviour thereby legitimise certain behaviours, creating standards that increase predictability and facilitate order in the relations between states. Through these processes, states construct international norms and institutions to govern the actions of members of the international political community

In this way, the rules and ideas of orderly conduct among states form social scaffolding that might be broadly equated with the roles of constitutions in liberal democratic societies. Organisations such as the United Nations, regimes and international law, are constructed by states to support and codify norms in order to enable them to 'tackle world issues' (Luard, 1990, p. 4). These institutions express shared beliefs and values that have been collectively constructed through the interrelations of states over time (Cameron, 2008). They outline and sanction principles that states understand as the basis of legitimate actions. International law and organisations, diplomatic relationships, treaties and understandings between governments all contribute to these processes.

Ideas and science

The unprecedented challenges arising from climate change render all members of the international political community more conscious of limits to their powers in relation to the material and environmental realities they confront. This realisation presents significant difficulties for states and the international political community because it necessitates new ideas about the roles, limits and capacities of states and their status as authorities. It also implicates states as perpetrators in human induced climate change and challenges their continuing rights to independence and autonomy.

The history of international environmental agreements reveals that incorporating professional expertise into government and intergovernment policy strengthens decision-making. Professional expertise enhances the flexibility of agreements and the adequacy of the policy evaluation mechanisms. As Schachter (1991, p. 457) observes, '[m]uch of the impetus for international legal restraints has come from outside national governments – from scientific communities, concerned publics and international organizations'. Schachter (1991, p. 459) believes the Principles of the United Nations Conference on the Human Environment held in Stockholm in 1972 effectively formed 'the starting point of international environmental law'. While international legal bodies, with the exception of the United Nations International Law Commission, hold only informal status, their rulings are nonetheless regarded by most states as authoritative.

Groups of scientific experts, non-governmental organisations (NGOs) and international organisations (IOs) are all, to varying degrees and in a variety of ways, knowledge and policy communities that influence the preferences of states and their perceived national and international interests. These are most effective when a clear correlation exists between the ideas they promote and those that states already hold. All states maintain national ratification processes that enable domestic political institutions to exert ultimate authority over internationally negotiated commitments, emphasising the undiminished nature of sovereign independence (Sprinz and Weiβ, 2001, p. 91).

Knowledge holding communities, which are sometimes called epistemic communities because of their shared political values and knowledge, are characterised by a common technical knowledge base and shared appreciation of its significance (Haas, 1997). These features enable members to develop a communal vision that contributes to states' abilities to identify potential problems, policy options and solutions (Young, 2002; Haas, 1997; Sebenius, 1992). These knowledge communities can cultivate particular roles within the international political community because they comprise informal networks of experts whose advice is considered credible and at arm's length from the political imperatives faced by governments.

Knowledge communities form through networks of professionals with recognized expertise and competence in particular domains and acquire authoritative claims to policy-relevant knowledge through their abilities to identify problems. Young (1989) identifies the quest for 'salient solutions' as an important aspect of their contributions toward international cooperation and finding durable and flexible solutions to global climate change problems. Scientific communities contribute knowledge and information and frequently also provide ongoing advice concerning potential solutions and strategies to policy-makers and international governance mechanisms (Haas, 1997). International awareness of the impending consequences of climate change has depended, to a greater or lesser extent, upon the roles of knowledge communities in dispersing information, shaping choices and developing common interests between individual (and competing) states.

In recent years, the public self-interest of key industrial actors has shifted from rejecting knowledge and projected consequences of climate change and water redistributions. Increasing numbers of them are now seeking a seat at the table in discussions, and pressuring governments on their own behalf. In part, this shift reflects the increased credibility of scientific projections and the attribution of knowledge status to scientific expectations of climate change consequences (Bodansky, 2001). While the status of projections remains contested in some quarters, over the last decade greater numbers of industrial actors have recognised that arguing over the details of possible consequences denies them opportunities for involvement in generating management and alleviation strategies. While industry groups are sometimes criticised by green actors for being motivated by a continuing quest for profit and ongoing livelihood, it is also possible to perceive their actions in terms of rationally driven enlightened self-interest (Pennington, 2001; Utting, 2000).

The politics of knowledge

Initially climate change was a subject of greater scientific rather than political debate and it struggled to gain acceptance beyond environmentally oriented organisations and institutions. As the precision of scientific research improved and uncertainties narrowed a better understanding of the role of anthropogenic emissions in global warming emerged. Consequently, scientists became more confident in their predictions and began to bring their concerns to the attention of governments, non-government organisations and the general public. As observed by Bodansky (2001, p. 27), '1988 marked a watershed in the emergence of...climate change...as an *intergovernmental* issue'. Over time, scientific research, the discovery of the hole in the Ozone layer and a series of severe weather events, including heat waves and drought in North America and the strongest hurricane ever recorded in the Western hemisphere, placed climate change at the forefront of public awareness and on the political agenda of a number of states (Bodansky, 2001, p. 27). The United Nations General Assembly (1988) confirmed this when it passed Resolution 43/53 which recognised 'that climate change affects humanity as a whole and should be confronted within a global framework so as to take into account the vital interests of all'.

Non-governmental organisations (NGOs) are knowledge communities comprised of non-state actors who do not seek to represent the political interests of states or governments, either formally or legally. These organisations emerge from civil society and seek to influence states' policies at domestic and international levels, functioning much like pressure groups. They are numerous and diverse, representing a multitude of views and interests on both sides of the debate about the seriousness and extent of global climate change. In this contested realm of ideas their impact cannot be denied and their contributions have been recognised by the international political community through their inclusion in a number of climate change negotiations. Many of the NGOs that are active in climate change debates have acquired ' "consultative status" with the United Nations...that allows them formal access to UN documents, negotiations and deliberations' (Raustiala, 2001, p. 97). These NGOs have become integral actors in global responses to climate change, providing conduits of ideas, knowledge and expertise between scientific and non-scientific communities.

Scientific knowledge has long been a platform upon which political notions and forms of organisation can be constructed. It provides motivation for moral action and strategies for ethical decision-making between states. It also underpins developments in international law, including some of the compliance and verification procedures that are associated with international agreements. In the context of climate change challenges, efforts to formulate legal definitions of such matters as 'environmental harm' and 'risk' have relied heavily upon scientific information, analysis and interpretation. Scientific knowledge has contributed to risk assessments, identification and promotion of steps to reduce risk or eliminate environmental problems, and evaluations of policy effectiveness. This has been crucial in enabling international political agreements and responses to global climate change.

An example of the influence of scientific knowledge can be found in the Montreal Protocol on Substances That Deplete the Ozone Layer which is an international treaty that entered into force on 1 January 1989. Under the terms of this protocol, signatories agreed to phase out the production of several substances responsible for depleting levels of atmospheric ozone. Scientific knowledge enabled states to modify the terms of the Montreal Protocol in response to expanding knowledge and new developments in science. International awareness of ozone depletion and the impact of greenhouse gases, which marked the beginnings of the international climate change crisis, was shaped both directly and indirectly by scientific expertise.

Without enough of the right kind of knowledge, states are doomed to make poor policy decisions. Their representatives rely upon knowledge as they negotiate, regulate, cooperate and review decisions. Knowledge holding communities promote consensual decision-making among states and enable them to identify incremental targets. In some situations, they also contribute to ensuring transparency in decision-making (Archibugi, 2001; Edmondson, 2001; Haas, 1997). Within international relations, these features of knowledge communities also contribute to trust-building between states, enabling them to form coalitions.

Assessments of environmental problems have depended upon knowledge communities that are able to transform uncertainties into positive contributions and provide spaces within which states might develop common interests. Drawing upon the work of Brennan and Buchanan, Young (1994) argues that uncertainty may assist in progressing bargaining. Although climate change responses remain unclear, cooperative and general principles for shared responsibility and cooperation have been and can be established. States and other actors are more likely to agree upon international mitigation and adaptation strategies when they share a common knowledge base, especially if they are making decisions while they remain uncertain about the particular benefits and outcomes (Young, 2002).

Ideas, authority and NGOs

States' interests cut across the boundaries of international relations and often emerge as expressions of domestic interests. Haas (1997, p. 4) argues, quite convincingly, that knowledge communities act as 'uncertainty reducers' supporting 'coordinated policy arrangements'. States are able to influence both processes and outcomes of negotiation and to express domestic interests more fully, not only through formal intergovernmental channels, but also through the less formal channels provided by knowledge communities.

The range of politically effective international actors that are able to influence the conduct of states has increased substantially as levels of interaction have increased due to economic, political and technological integration. Climate change issues cut across and impact upon many of the domestic and international activities of states and therefore attract the involvement of great numbers of NGOs. The authority of states has not been diminished by this development, since they are free to either accept or reject advice, but the influence of these new voices upon the direction of states' policies is significant (see Chapter 4).

The NGOs that are active in climate change issues pursue dual strategies of exerting social and political influence. They raise and maintain public awareness and interest in climate change issues through media engagement and fund raising efforts to sustain their activities. Many of them have become household names such as Greenpeace and the World Wildlife Fund (WWF). Their ideas about climate change and its various effects upon human societies and geophysical systems contribute to public perceptions of what constitutes appropriate governmental and economic activities. In this way they seek to influence the conduct of individuals and to create public pressure and expectations that government policy-makers cannot ignore. At the level of influencing state activities, NGOs provide important sources of relatively inexpensive expertise and information. Their knowledge and influence is important for all states, especially those that may not be able to resource their own sources of knowledge, scientific and technical expertise (Raustiala, 2001, pp. 104–105). They have also 'helped to "multilaterize" [sic] information' about international climate-related actions by analysing 'what governments have claimed to do, what they have actually done, and what is likely in the future' (Raustiala, 2001, p. 108).

Importantly, NGOs do not represent a monolithic bloc of unified interests that influence or coerce the activities of states in any one direction. Their interests are as diverse as those of the states they petition to reduce greenhouse gas emissions and respond to changing rainfall patterns and other climate change impacts. They represent, among others, the interests of environmental groups who believe that the direction of human development has adverse effects upon the ecosphere and requires reconsideration. They also represent business and industry groups who disseminate contending science and viewpoints to promote their own interests and responses to global climate change.

Other less obvious sectors of activity, such as the legal and medical fraternities, have also become involved in the climate change debate and sought to influence the development of international responses (Parry et al., 2007). As observed by Raustiala (2001, pp. 106–107):

> One of the most prominent examples is the relation between the London-based Foundation of International Environmental Law and Development (FIELD) and the Alliance of Small Island States (AOSIS). Members of FIELD, mostly international lawyers, consulted extensively with members of AOSIS, appeared on their delegations, and at times acted as the delegation of certain AOSIS members. The tiny member governments of AOSIS, often lacking much indigenous expertise about climate change and the policy possibilities, became a more powerful negotiating force in conjunction with FIELD.

Similarly, the 1997 publication of an *International Physicians Letter on Global Climate Change and Human Health* by the Physicians for Social Responsibility sought to influence and inform international policy negotiations (Raustiala, 2001, p. 111). In both examples, NGOs utilised perceptions of their political neutrality or pluralism and expertise to position their own ideas in the international policy process.

Ideas about the extent, urgency and nature of global climate change impacts affect the prospects of effective international responses (Lieserowicz, 2006). They impact upon social, political and economic decisions and activities among all actors in international and domestic decision-making contexts. Achieving political action depends upon a broad acceptance of ideas. The resolve to apply the resources necessary to address the effects and causes of climate change are likely to emerge only when there is widespread acceptance and political will directed towards acting upon common perceptions. As such, a community of concern needs to be created before significant responses can be orchestrated internationally. In recent years, NGOs have demonstrated their abilities to play integral roles in international negotiations, acting as 'agents of social change' through their contributions to knowledge and efforts to influence individual and collective political responses (Pincen and Finger, 1994, pp. 64–65).

Conclusion

Policy-makers rely upon scientific predictions and specific knowledge to identify, understand and seek solutions to global climate change effects. Common values and interests are crucial, because the international community is vulnerable to the magnitude of social, political and economic consequences that climate change will produce. Knowledge communities, as non-state actors, play important roles in developing common visions among states (and NGOs) and can support their pursuit of common objectives. Knowledge communities can thereby reduce the dominance of strong governments. By providing information to many states, inter-governmental organisations and members of the public, they ensure knowledge is shared, providing a common basis for decision-making that might enhance informed policy-making.

Effective international organisations rely upon sustained cooperation between parties who identify mutual or complementary interests (Young, 1997; Haas et al., 1993). Implementing international climate change policies inevitably depends upon international organisations, the agreements states and other actors achieve, and the structures and organisations they create and actively maintain. Mutual interest and the political will to resolve conflict through negotiation and bargaining, and to pursue the targets or objectives embodied in international governance mechanisms, are essential to the creation of agreements and the policy implementing capacities of international organisations (Edmondson, 2011).

In the context of international environmental organisations, a partial separation is frequently maintained between scientific research and policy-making (Edmondson, 2001). This enables research parameters to remain largely separated from formal policy discussions. International scientific groups actively engage in complex negotiations to communicate knowledge and values to policy-makers, but are not active policy-makers in their own right. Their influence lies in their research and the knowledge, including expectations and predictive capacities, they generate. They establish early commitment among parties and play roles in working with states to determine the costs and relative merits of possible responses. In this way they contribute to trust building among participants.

The degree to which knowledge communities have been successful in implanting ideas about climate change in the popular consciousness was demonstrated by a 2007 BBC World Service poll that found:

> 65 per cent of the 22,000 people polled in 21 countries said there was a need 'to take major steps very soon' ... The poll showed 9 out of 10 people wanted some action on climate change, and 79 per cent said human activity was contributing significantly to the problem that scientists say would cause major hardship worldwide.
> (*Herald Sun*, 26 September 2007, p. 33)

Contributing to this are efforts by individual activists, such as 2007 Nobel Peace Prize co-recipient Al Gore, whose climate change film *An Inconvenient Truth* has been screened globally. Evidence of the extent to which such ideas have permeated the policy-making considerations of states is reflected in comments by Ban Ki-Moon, Secretary-General of the United Nations, in his address to the High Level Meeting on Climate Change in New York on 24 September 2007: 'the nature and magnitude of the [climate change] challenge' can only be addressed by 'contributions from all countries and all sectors of society' because 'national action alone is insufficient .... That is why we need to confront climate change within a global framework, one that guarantees the highest level of international cooperation'. This confirms the important role of ideas in bringing the interests and preferences of states into closer alignment such that shared concerns provide a platform for an effective and cooperative international response.

10
Rights, Responsibilities and Sovereignty

Introduction

Climate change is a problem of extraordinary scale and constitutes one of the most significant challenges ever to confront the modern global community. Like the potential of global nuclear war, climate change threatens the continuation of modern societies. Prevailing scientific consensus, even at their most modest, is that it will fundamentally alter the political, social and economic activities of all societies. These effects call into question many of the development practices that have produced and sustained the contemporary world. In so doing, they undermine certainty in future industrialisation and raise doubts concerning current notions of progress.

Reconsidering sovereignty and the nature of states is necessary for developing effective responses to climate change because the established structures and capacities of sovereign states provide the basis for orderly conduct in the international community. Sovereign states constitute the principal forms of political association and may lose their abilities to provide physical and social security for their people and struggle to retain their legislative, regulatory and resource management capacities (Edmondson, 2011; Kütting, 2000). Although some states may be in a position to exercise leadership in responding to climate change, it seems inevitable that others will lack adaptability, struggle to demonstrate timely responsiveness to information, or wilfully resist attempts to pursue responses under the leadership of others (Wilson, 2008, p. 10). It is therefore necessary to consider the roles that might be played by states that are unable or unwilling (for reasons of ideology or capacity) to make meaningful contributions to international climate change responses.

In the 21st century, the nature of sovereignty will inevitably be affected as climate change impacts upon the territorial possessions of states and they attempt to reduce greenhouse gases and other industrial emissions by altering their economic activities and structures. It will become necessary to reconsider the political nature of states in terms of their rights to claim exclusive authority to regulate the use of resources within their territories. Their capacities to claim rights and responsibilities on the basis that they represent their citizens' interests will be altered by global climate change (Young, 1994). As some states lose territory to rising sea levels and others lose rivers through changed rainfall patterns, they will expect to find support from their peers as they attempt to provide ongoing security for their citizens.

While the anticipated extremes of global climate change are still being identified, there is overwhelming evidence that its effects threaten the continuation of current forms of societies. The largest modern cities rely upon high inputs of energy and food supplies from remote locations and are likely to become unsustainable (Roaf et al., 2005). These predictions are difficult for many to imagine from the comfort of their technically advanced middle-class circumstances, or from the discomfort of poverty. Not surprisingly their accuracy and validity continue to be debated. While a vocal minority of scientists and other experts argue over the timeframes of particular climate change impacts, or the accuracy of current predictive models, other members of societies remain distrustful of such predictions because effective responses rely upon sweeping changes to the foundations of their daily lives.

In spite of these complexities, the best efforts of scientists over several decades suggest that climate change is real. Contesting their views means questioning both the integrity of the scientific method and the utility of specialisation upon which much of the industrial age has been based. This challenges some of the core features of modern societies which have long believed that new scientific information and knowledge will invariably improve their life expectancies and comfort levels. Nonetheless, most scientists now agree that global climate change cannot be prevented from reducing the availability of productive land and the number and distribution of species. Most now also agree that it results from the cumulative effects of human activities over several centuries (Intergovernmental Panel on Climate Change, 2007b).

Some scientists and environmentalists have long perceived the combined effects of industrialisation, urbanisation and more recently also globalisation as dangers to long-term planetary survival (Schumacher, 1974; Toffler, 1971). Many of the recent climate change predictions have focused upon the impacts of specific pollutants or other forms of environmental destruction. To some extent, focusing upon specific pollutants made it easier for large numbers of states to agree upon international targets and standards, such as in the 1989 Montreal and 2005 Kyoto Protocols. However, these political dynamics contributed to international responses that were restricted to regulating certain products, such as ozone-depleting substances and the production of greenhouse gases. Rather than enabling governments, producers and other economic actors to come to terms with the more complex and challenging problems arising from global climate change, international agreements such as these enabled political leaders to avoid addressing more difficult climate change adaptation and mitigation strategies.

Although pockets of scientific and knowledge-related debates continue, there is now a preponderance of scientific agreement. The scale and rapidity of global climate change effects are generally accepted as long-term consequences of industrialisation and patterns of modern life. Among its political impacts will be revisions to the rights assumed by states and their abilities to maintain central roles as political authorities (Vogler, 2005; Kütting, 2000). Climate change thereby presents additional political challenges in terms of the range of social interests that can be accommodated and afforded priority.

Accepting climate change

Throughout the modern international community, sovereign states have outlined and utilised common rules of mutual recognition that have allowed them to relate to each other as equals. This has enabled states to act autonomously, holding independent authority and regulatory capacities over domestic issues within their territory (Kegley and Raymond, 2005, p. 47). In the 20th century, independent or unilateral actions were often justified by states through the regularly invoked principle of sovereignty which reaffirmed their autonomous authority. These foundations of international political behaviour exacerbate the challenges of achieving collective agreements among states and increase the importance of effective political leadership.

The difficulties in identifying links between specific climate change effects and natural disasters increase the political difficulties experienced by states, especially those based upon forms of representative government. Developed, industrialised, liberal democracies confront particular political difficulties in accepting predictions about the likely long-term effects of climate change. First, the timeline for consequences exceeds the cycles of public imagination and most accepted economic and political models. Even as scientists and environmental activists argue for immediate action it has largely been impossible to attribute specific natural disasters or environmental problems (such as drought or bushfires) to climate change (United Nations Development Programme, 2008). Nonetheless, exceptions have occurred, most notably when they have contributed to or triggered violent conflict. For instance, drought in Darfur (and conflict arising from it) has been identified as a consequence of desertification (Flannery, 2005). Also, the increasing incidences of hurricanes affecting North America over the last decade were attributed, in January 2008, to an increase in the surface temperature of the Atlantic Ocean (Climate Science Watch, 2008).

If the anticipated consequences of climate change are to be averted or reduced, efforts to address them must commence before they impact upon daily lives. This presents an additional political challenge because voters tend to react to immediately visible triggers. Failure to address climate change in a timely manner by imposing sustainable burdens upon contemporary lifestyles will result in imposing impossible burdens upon future generations. This creates a disjuncture between views of 'urgency' concerning impending consequences and other more readily identified 'immediate' imperatives such as avoiding interest rate increases or the closure of major production facilities.

International policy-makers are rendered powerless by their vulnerability to electoral sensibilities that remain insensitive or unwilling to accept the costs of climate change mitigation or adaptation strategies. Political and business leaders find themselves in a position where they need to lead rather than respond to public opinion. The decisions they are required to make, and the responsibilities they must assume in order to avoid the worst case climate change scenarios, will negatively impact upon daily activities in all societies. Their task necessarily entails new responsibilities in public education and new linkages between the economic and political spheres of many societies.

A second impediment to responding to the predicted effects of global climate change centres upon debates about the nature and extent of human contributions. While changes in climate patterns are widely acknowledged and observed, debates continue about their causes. This is a particularly slippery component of international policy-making because changes in climate have occurred throughout world history (Linden, 2006; Flannery, 2005). Some level of climate change can be explained as a natural occurrence which adds complexity to international negotiations and initiatives for managing its impacts. As states and key economic actors confront climate change as a threat to economic prosperity, they seek to apportion responsibility and distribute response costs by identifying the relative balance between naturally occurring phenomena and human contributions.

Individual memories tend to be highly localised and lack the longitude to permit reliable evaluations of climatic changes. Institutional memories, such as those of states, tend to be less localised and are measured in hundreds of years. Nonetheless, they are dwarfed by global climate cycles that are measured in thousands of years. Hence, a cautionary approach appears justifiable to sceptics who emphasise the fallibility and constructed nature of scientific specialisation and knowledge.

As scientific and political debates have progressed, two causal chains have been introduced for public consideration. The first is that the global climate changes over time and these changes have historically pushed some human societies beyond their abilities to adapt (Linden, 2006, pp. 26–27). The second causal chain requires us to accept that human activities are changing the world climate (United Nations Development Programme, 2008). This chain identifies technological activities as directly contributing to changes in global climate and argues that the nature of these changes will be perilously disruptive.

A third complication to accepting current predictions of the nature and effects of climate change rests upon a general perception of climate as a natural and immutable backdrop to human activities (Linden, 2006). This is an important factor in shaping the political difficulties associated with addressing climate change consequences. Societies are accustomed to natural environments that provide resources and seasons with a level of certainty, albeit with periods of localised disruption. The idea that permanent disruptions are occurring is therefore challenging.

Contested rights to authority

According to common convention, a state consists of four integral components – territory, population, government and sovereignty (Morgenthau, 2005; Waltz, 1979). First, a state is a territorial construct with clear political boundaries in the form of borders that are widely recognised by other states and actors. At this most basic level, states are synonymous with what might, in other contexts, be called countries. These terms, however, are not precisely interchangeable and the difference is in their political status. Countries are merely geographic entities and cannot be presumed to hold recognised jurisdictional authority, while states are characterised by their decision-making and law enforcement capacities.

States arise from political ideas and values that perceive a foundational relationship between government authority, rule of law and territorial identity. They pursue particular interests and can be thought of as having specific goals, desires, concerns and fears. In contrast, countries are passive geographic entities that are acted upon and within. These differences are important for global climate change responses because states are actively constructed political entities that are engaged in a perpetual, and often bewildering, array of activities and associations (Oye, 2005; Wendt, 2005; Keohane and Nye, 2001). Countries are often socially linked because of their geographic proximity and many are also economically linked by trade. However, states seek to preserve their authoritative identities by fostering common recognition of their political authority.

Sovereignty provides states with unique political authority. In addition to territory, a state has a recognised population composed of one or more ethnic identities who conceive of themselves as holding a broader state-based identity (Farrands, 1996). According to the ideals of sovereignty, states create a unifying national identity among the population within their territory. This national identity derives from the construction and maintenance of a variety of state activities and apparatus, such as flags, anthems, shared values and borders (Anderson, 1991). In consolidating their national identity, a state differentiates its population from others who live outside of its boundaries and have different allegiances (Camilleri et al., 2000).

A recognised government that answers to no higher authority and exerts control over both territory and population is a key component of statehood (Morgenthau, 2005; Bull, 2002; Waltz, 1979). These principles enable governments to be considered as legitimate representatives of the state in their internal and external activities. They exert authority over the use of territorial resources and the behaviour of their populations. However, territory and population alone do not make a state, and where this is the case intra-state conflict is often apparent. In such situations, the absence of a common identity or allegiance to a common authority promotes conflict and activities by separatist groups (Jackson and Sorensen, 2007; Krause and Renwick, 1996). These dynamics will prove increasingly problematic as water, energy and agricultural land shortages increase in coming years.

Since the middle of the 20th century there has been a steady shift away from the principle that state sovereignty should be protected irrespective of its consequences for individuals, groups and organisations. Contributing to this has been international acceptance of the principle of collective security and the democratising project of Western states which produced an array of human rights instruments. In both regional and international politics, sovereignty is now taken to imply respect for an extensive range of human, resource and labour rights. These have created a new set of ordering political principles that potentially limit, but do not erode, state power as an essential component of sovereign authority.

Additional political difficulty lies in debates between and among states about who holds responsibilities for progressing climate change responses and who should bear the costs of their implementation. Little consensus exists about the relative distribution of burdens and responsibilities among states and economic actors based upon their past and present contributions to climate change (Roberts and Parks, 2007). In these there exists a crude division between developing and developed states. The developing states see themselves as further disadvantaged and marginalised by new international efforts to regulate the technologies and industrial practices they desire. They believe the greater burden should be shouldered by developed states. Not surprisingly, the developed states who have been major contributors of greenhouse gases see the situation differently. They argue that climate change is a global and collective problem that can only be effectively addressed with all states working and making sacrifices together (Eakin et al., 2009; United Nations Development Programme, 2008).

Hope and ingenuity

Difficulties with accepting predictions about the likely long-term effects of climate change do not, however, constitute a rational justification for a lack of action and deferred responsibility. While the long timelines involved make the need for immediate action difficult to accept, it should not be beyond the understanding of the modern global community. If debates about whether the effects of climate change are already clearly visible are put aside, it becomes less contentious to imagine daily impacts and irreversible effects that will be readily visible in 50 years. This time frame is broadly equivalent to that between the Wright brothers' first flight and the moon landings. Doubtless, the observers of early flight gave little thought to such distant and heady achievements but the modern world now finds itself requiring a similar scale of imagination.

Fortunately, advances in knowledge and technologies that allow such achievements also support efforts to ameliorate the worst consequences of global climate change. Technologies and expertise provide the tools with which to accurately record thousands of years of the past and these same tools provide the knowledge for effective future planning. Debates concerning whether or not human activities cause or contribute to climate change are merely a distraction that flatters our egocentricity. According to the best predictions, the global climate is changing and it is possible to mediate some of the most disruptive effects this will have on human societies in the future (Eakin et al., 2009; United Nations Development Programme, 2008; Metz et al., 2007).

Continuing to argue about the accuracy of particular scientific interpretations about what is occurring within the natural environment is a short-sighted and largely esoteric debate over the relative merits of different research methods. These debates then spill over into related discussions concerning how communities and their political authorities should respond to climate change impacts. In this context, they become as dangerous to future generations and progress as climate change itself.

Ongoing displays of faith in human ingenuity to provide timely responses and solutions to climate change rest upon long-term political will and an acceptance of the progressive nature of scientific knowledge. One, without the other, threatens to unravel what has been a dynamic and advantageous relationship. Without political will to invest in scientific advances many of the developments of the modern world would not have materialised. Likewise, without the predictive and productive capacities of modern science many of the political aspirations of the contemporary world would have amounted to much less (de Coninck et al., 2007).

De-coupling authority and territory

One of the features that distinguish sovereign states from other political entities is their rights to claim and delineate territory. It is on this basis that states then claim to exercise exclusive jurisdiction over their territory, assert their rights to determine borders and exercise control over those borders. In this way, sovereignty enables states to 'distinguish a specific political community – the inside – from all others – the outside' (Doty, 1996, p. 122).

Climate change challenges sovereignty by blurring the internal (domestic) and external (international) functions of governmental authority among states. What constitutes independent authority is changing along with the areas over which states exercise authority as increasingly what a state undertakes within its own territory has global 'real world' impacts. Arising from this will be renewed concern about the prerogatives of statehood and the ethical exercise of state authority. As a consequence, ideas about sovereignty will begin to change. If the international community is to mount a coordinated and timely response to climate change and its effects it will be necessary to change the functions of states as the foundational units of human communities.

The conduct of sovereign states has been shaped by their prerogatives as sources of authority. Their functional attributes and beliefs in their institutional importance emerged against a particular appreciation of the relationship between humanity and the natural environment. States were derived and developed according to a particular understanding of the relationship between human societies, the resources and natural environments in which they existed (Hashmi, 1997). The international political community gave states additional prerogatives over time to facilitate and regulate their relationships.

Climate change, however, is beyond the territorial experiences of states. It produces new interdependencies and interactions that have little concern for the location of borders or cartographic position (Young, 1994). Many of the interactions that states initiate and respond to in their dealings with each other (such as trade, communication and management of migration) have been, and continue to be, shaped by climatic processes. At the same time, these interactions encourage a confluence of ideas and information between states and a diverse range of non-state actors (see Chapter 4 for further discussion of the importance of non-state actors). These are embedded in a climate driven natural environment that now demands new awareness of, and responses to, unfolding climate change impacts. Governing and managing the international political community to ensure continuing habitability and sustainability of human societies requires acceptance of these new political realities.

International order and stability

The climate is changing the world in which sovereign states are situated. At the same time, there are increasing numbers of, and interactions between, non-state actors. As a result, states are no longer isolated from overlapping networks of interdependence across their political, economic and social spheres. The irrelevance of borders also contributes to a weakening of the theoretical inside/outside dichotomy as states' internal behaviour is no longer private (Farrands, 1996). How states determine their national interests carries over into the international sphere because domestic policies concerning climate change affect can have global ramifications.

Increased communication, migration and trade have added layers of complexity to state relations that renders traditional thinking about their rights and authority too simplistic. Statehood, in part, relies upon a demonstration that national actions comply with universal values. It is possible to imagine a not-too-distant future in which such measures might encompass climate change issues. In such a future, if states did not contribute positively to alleviating climate change, then their status as legitimate actors could be called into question.

Global telecommunications are reorienting states' interests and broader political structures (Held et al., 1999). With information about the internal activities of states more accessible, there is greater pressure on them to justify their domestic policies. States cannot be entirely autonomous, because they are part of a community containing ubiquitous interests in self-preservation. In some cases, national goals are subordinated in preference for shared goals, emboldening the presence of co-existing internal and external state responsibilities. Internal responsibilities also contribute to the obligations of international actors.

The principle of sovereignty enables states to claim rights (Bonanate, 1995). However, the political implications of global climate change reinforce an alternative view that sovereign states have primary duties towards others (Appiah, 2006). Presently, legitimacy is determined by the levels and manner of participation and interests pursued by states. International treaties, for instance, symbolise important legal frameworks for appropriate actions. States that abide by international laws, and international environmental laws in particular, demonstrate concern for the shared interests of the global community and this is increasingly important for sovereign legitimacy (Held, 2010). Extending and maintaining perceptions of legitimacy relies on states being accepted as members of an international community that is increasingly centred upon similar interests.

International cooperation and international law are built upon negotiated compromise in the competitive activities between states. These are two of the most important tools available for addressing global climate change as it disrupts societies. However, both international cooperation and law rely upon ideas, concepts and values that derive from a desire to impose order over human interactions through mediation and compromise. These are problematic tools for addressing global climate change because they are located in the realm of political ideas and organisation, while climate change is largely a matter of physical and material change. By their very nature these tools are limited in their abilities to tackle a physical phenomenon. The international community cannot avoid the effects of climate change by negotiating responses among themselves. Nevertheless, strategies to address global climate change will only begin to emerge when human communities accept collective responsibility for finding solutions (United Nations Development Programme, 2008).

The use and ownership of resources lies at the heart of contemporary government. For modern states, everything – the provision of food, shelter, employment for the population, production and the balance of trade – depends upon the availability and management of an extensive list of resources. Maintaining domestic and international political stability through sovereignty, security and economic growth are intimately aligned with resource use. States then struggle to respond positively to threats to their resource supplies. They experience policy paralysis that produces 'wait and see' approaches (Yohe et al., 2004; Lind, 1995).

During the 20th century, creating and maintaining fixed, and recognised political borders between states, consolidated their sovereign authority and progressively reinforced states' expectations of territorial security (Morgenthau, 2005; Keohane and Nye, 2001). This was achieved because most states and the international political community accepted sovereignty as central to orderly societies and promoted the political authority of states above other forms of territorial or social authority (Murphy, 1996). This belief in the overarching importance of political authority then enabled the creation and maintenance of borders and ensured that diverse states have recognised and attributed basic similarities.

Effective responses to climate change rely upon accumulated knowledge about state sovereignty to ensure continuing order (Edmondson, 2009). By premising sovereign authority upon independent law making within specified territorial boundaries, the international political community historically relied upon the spaces between states to create order. In recent decades these spaces have become reduced and crowded as interactions among states have increased and intensified. The contemporary international political and economic systems have, quite deliberately, brought communities into closer and more sustained contact (Young, 2002). Climate change and its increasingly evident effects further complicate these relationships and increase sources of potential conflict. The uneven distribution of global climate change impacts have the potential to exacerbate existing tensions as states become increasingly concerned with the internal activities of other members of the international community.

States' rights

States claim territory and exercise authority by enacting and implementing laws, and recognising the rights of other states (Edmondson and Levy, 2008). This relationship between territory–authority–sovereignty establishes states' rights and underpins their abilities to exercise jurisdiction over defined, and recognised, territory. In very practical ways, states claim and exercise their rights of sovereignty through their supreme authority within a jurisdiction (Herz, 1957). This is expressed as an ability to create laws within territorial limits that provide legal boundaries and separate different sovereign entities.

One of the political attributes of sovereignty is that it provides theoretical equality among states by which differences or conflicts can be overlooked or overcome. Under conditions of mutual trust, it enables states to negotiate, bargain and act collectively (Edmondson, 2011). Sovereignty is highly valued because it provides states with autonomy and supports their internally focused decision-making (Held, 2010). However, the political, economic and social challenges arising from climate change are likely to disrupt these mechanisms, leaving the international community vulnerable to uncertain political structures and potential disorder.

A key weakness in sovereignty as an organising principle for relations between states is its inability to create functional or real equality. Instead, it provides a façade through which unequal forms of 'equality' are maintained and through which the independence promised by sovereignty has become an increasingly powerful ideal (Edmondson and Levy, 2008). As such, 'equality among states' has been promoted, supported and reinforced by international collective decision-making (Hoffman, 1997). The localised impacts of climate change will highlight the extent to which this theoretical legal equality contributes little to the practical, functional dimensions of governmental authority as not all states exert supreme authority over the same range of issues (United Nations Development Programme, 2008; Young, 2002). Contemporary states individually determine for themselves those areas of political activity in which they choose to exert absolute and supreme authority, and those areas in which they are happy to share or delegate authority (Edmondson and Levy, 2008).

Over the last 300 years, sovereignty has evolved, transforming the prerogatives that states enjoy (Philpott, 1997). For modern states, sovereignty no longer has to be absolute in all areas for them to remain primary and supreme sources of authority within territories (Philpott, 1997). Supreme authority is not measured in terms of 'absoluteness', but rather 'the scope of affairs over which a sovereign body governs within a particular territory' (Philpott, 1997, p. 19). A state can surrender areas of sovereign interest and concern to international law, or internationally legitimate institutions, and, to date, doing so has been crucial in enabling climate-related agreements and structures (Hoffman, 1997; Henkin, 1989).

States' responsibilities

Although non-state actors, such as multinational corporations and non-governmental organisations, are readily identified as contributors to knowledge and political interests, they are not, ultimately, responsible for action or inaction by and among states. When issues arise impacting upon the territory or population of multiple states the international community expects the sovereign states to deal with one another rather than to attempt to resolve the matter unilaterally in their own best interests (Fowler and Bunck, 1995). This is one of the reasons there is an expectation that climate change issues will be addressed internationally rather than by individual or discrete groups of states.

Within the international political community, states are also held accountable by their peers. They are required to respect borders and uphold the principles of collective security. The key distinction between states and other actors lies in states holding 'political responsibility for governing, defending and promoting the welfare of a human community' by virtue of their sovereign nature (Fowler and Bunck, 1995, p. 12). However, non-state actors also fulfil a range of requirements in the modern integrated political environment and are often crucial in preserving orderly conduct among states, even though they remain dependent upon the validation of sovereign states (Hoffman, 1997).

The sovereignty of states continues to provide an important source of stability despite the fact there are readily identifiable transnational issues that defy individual state resolution and which the international community seems reluctant or unable to confront (Holsti, 1995). It is easy to be critical of international inaction on climate change issues, such as the slow pace at which the Kyoto Protocol was ratified, and the persistent refusal of some states to join it. However, given the diverse nature of states and their competitive pursuit of resources and security, it is remarkable that international political agreements and structures are achieved. Political fragmentation, competition and conflict are symptomatic of an inability of the international community to address problems that 'rarely correspond with state boundaries' (Murphy, 1996, p. 108).

Claims about the decline or erosion of sovereignty suggest states are less responsible and hence less accountable for transnational issues. From the viewpoint of seeking to maintain economic prosperity this may appear desirable in terms of the movement of capital (Ohmae, 1995). However, the abrogation of state accountability over transnational environmental issues is neither progressive nor a sound basis for addressing global climate change. The required innovations to mediate and adapt to the effects of climate change can only be achieved in an orderly manner by states, secure in their sovereignty, embracing additional responsibilities (Edmondson, 2011). Increasingly, states will be required to collaborate and cooperate for the provision of ecological security.

The ethics of sovereignty

Sovereignty requires states to safeguard the welfare of their citizens. Similarly, sovereignty impels states to seek prosperity by guaranteeing resource and territorial security alongside predictable and accountable government. It also carries ethical elements that commit states to maintain and protect their territory, recognise the authority of other states, and to ensure their survival as authoritative political entities through time (Appiah, 2006). Mutual recognition of sovereignty is part of the accepted behaviour and business of states in their routine international dealings (Edmondson and Levy, 2008).

Sovereignty ensures that major international political issues, such as human rights or environmental protection, can be dealt with effectively without undermining the domestic authority, decision-making or juridical capacities of states. It creates both the precondition upon which states hold, assess and identify their national interests, and also the imperatives that drive their assessments and efforts to alleviate common problems (Paterson, 2000). Nonetheless, domestic politics are not altogether subservient to international political decision-making and the implementation of international policy goals. Sovereignty is also both the foundation and driving imperative of international management mechanisms that seek to regulate the use and exploitation of global commons resources.

In the latter half of the 20th century, more globalised relations and the rise of new political voices highlighted the ways in which states have been granted both 'rights and duties' in exercising domestic political authority while their external ethical responsibilities have often been overlooked (Bonanate, 1995, p. 9). Global climate change threats to economic progress mean that states' responsibilities for protecting the environment and ensuring their people have access to lifestyle sustaining resources can no longer be denied. In the early 21st century, states have experienced increasing pressure to demonstrate that their sovereignty is grounded in international legitimacy. This can be demonstrated through their willingness to accept responsibilities for responding to global climate change. A state that does not accept responsibility for international commons preservation and seek to alleviate the worst disordering effects of global climate change, might expect to find its legitimacy questioned. As states are becoming less isolated and less autonomous, it is their responsibilities, duties and obligations that bind them to interdependent activities and decisions.

Sovereignty is changing due to the shift away from individual state goals and interests in favour of increased consideration for the international political community. Ideas of states and sovereignty are in flux because their meaning is 'not fixed...but...undergoing change and transformation' (Biersteker and Weber, 1996, p. 14). As global climate change impacts upon people, territories, economic and agricultural production and forms of human societies, the principle of sovereignty will need to display new levels of flexibility. This is especially the case if sovereign states are to remain the central structures and sites of political activities in the international political community in the 21st century.

Conclusion

The difficulties in identifying links between specific climate change effects and natural disasters increase the political difficulties experienced by states with representative governments. Attracting and sustaining popular support for initiatives to address and reduce climate change impacts and triggers have proven difficult because of the time scales involved and the immediate costs that are imposed. Political will, initiative and policy cycles ebb and flow with electoral cycles as voters react to policies and personalities. Representative governments tend to operate according to electoral horizons of two, four or six years with a need to demonstrate good economic management within that period.

Sovereignty underpins the capacities, responsibilities, rights and obligations of states (Edmondson and Levy, 2008). It evolves when the prerogatives that states enjoy are transformed by changes in their abilities to exert effective authority and maintain structures of mutual recognition. The capacities of states have, to a certain extent, been historically founded upon their natural endowment of resources. More recently, the international political economy, technology and political affiliations have allowed states to overcome the tyranny of distance, accessing resources and markets beyond their own borders. Global climate change is, however, altering the composition of the international community, the characteristics of states and societies. In this regard, climate change poses genuine challenges to current ideas about how societies are politically organised and represented. It further impacts upon the ways they seek to progress their interests, including pursuing economic prosperity and rights to political participation and representation.

Global climate change creates real-world political challenges and theoretical political challenges that then generate real-world consequences (Pennington, 2001). Ideas about independent sovereign states and the practices of the international political economy constitute a dangerous nexus. The competitive and territorially based practices of states were once entirely appropriate for creating and maintaining their sovereignty and supported orderly relationships between communities. In the 21st century, however, they have become institutionalised impediments to effective international climate change mitigation and adaptation strategies.

The characteristics of political authority that have been cultivated among states have increased their drives for economic prosperity and industrial progress, making it more difficult for them to accept and recognise their inherent reliance upon nature (Young, 2002; Shrader-Frechette, 1998). In important respects, the economy is a wholly owned subsidiary of the environment and societies need to acknowledge their dependence upon the climatically controlled environment in which they are constructed (Linden, 2006). As more sophisticated understandings of the natural environment have developed, along with an appreciation of our capacities to modify and manipulate it, there is greater appreciation of humanity's need to accept more responsible stewardship (Dryzek and Schlossberg, 2005; Eckersley, 2004).

Addressing climate change requires the autonomy of statehood to be mediated by internationally articulated and environmentally responsible parameters. Increased accountability, through either new and strengthened mechanisms for global governance or the popular expectations of informed citizens, will transform the prerogatives and privileges of statehood. In doing so, it will cause a further evolution in what is understood as state sovereignty. The sovereign state can no longer be considered a bounded space, because states can no longer act irrespective of global interests. While there remains a place for autonomous actors in the international community, this autonomy can only be considered limited because climate change mitigation and adaptation strategies rely upon coordinated responses. Achieving predictable and orderly cooperation depends, in turn, upon effective leadership within the international political community and the states that constitute its primary components.

11
Identity, Ethics, Security and Order

Introduction

Climate change impacts present a series of escalating political ramifications for states and the broader array of actors who comprise the contemporary world. Responding to impacts such as rising atmospheric temperatures and sea levels, loss of agriculturally productive land, redistributions of water resources, increased storms and severe fires all demand new forms of action from political actors and organisations. These changes impose new responsibilities upon political, economic and social actors, including those who remain reluctant to accept new levels of authority and collective action. The Copenhagen Summit 2009 demonstrated the extent to which government leaders continue to clutch at historically based concepts of sovereign authority, prosperity, harmony, order and international influence. States continue to hope that their interdependent geophysical systems will prove amenable to independent management. Indeed, the limited outcomes of the Copenhagen Climate Summit suggest that states regard their political independence as more important than addressing global climate change.

In the 21st century, climate change issues are highlighting environmental, resource and political interdependencies among states and overlapping interests between various economic and social actors. These changing political dynamics are revealing their common reliance upon collective responsibilities (Edmondson, 2011). Political leadership must now take account of inherent interdependencies and their implications for the possible lifestyles and forms of political association that might be sustained (Biermann and Bauer, 2005; Archibugi, 2001; Postiglione, 2001). The ability to exercise reason, hope and confidence is central to the nature of states – and these features arise from their moral and legal standing as sovereign authorities (Bonanate, 1995).

As states seek to secure their peoples and territories, they necessarily also confront a series of tensions between their hopes of prosperity, harmony, order and international influence. States attach tremendous importance to these considerations and global climate change is making them more sensitive to their inabilities to control the forces that underpin their political endeavours. Additionally, many states remain uncomfortable with dawning realisations that their prospects of maintaining territories, exercising authority and protecting their peoples might be diminishing rather than expanding. Shifting rainfall patterns, the extinctions of species and the spread of diseases produce a series of new demands upon the policy- and law-making capacities of states, and these are yet to be universally recognised as fundamental security concerns (Barnett, 2001).

International order is linked with the authority of states and relations between them and the self-perpetuating norms of international order reinforce states as primary political actors. International order arises from and establishes conditions within which states subsequently agree to compromise and reconcile their competing interests in exchange for the benefits of a relatively stable political environment (Morgenthau, 1973). However, maintaining order while the environmental crisis is addressed requires every state to accept compromise. They must reconsider their competing interests with every other state in exchange for the benefits of a relatively stabilised natural environment. States, at one very fundamental level however, were created to permit competition and ensure the interests of diverse national identities were not compromised for the interests of a broader community (Morgenthau, 1973).

Global climate change is forcing many people to amend their economic enterprises, relocate their homes, reduce their rates of fossil fuel energy consumption and re-evaluate human relationships and interactions with the natural environment (United Nations Development Programme, 2008). In spite of these observable climate change impacts, the Copenhagen talks failed to produce substantive international agreements concerning emissions targets or globally accepted approaches to industrial and other adaptation strategies. The Copenhagen Climate Summit made clear that new political structures supported by patterns and perceptions of authority will be essential components of effective international mitigation and adaptation strategies.

If new international discussions and bargains ultimately emerge these are unlikely to result in universal agreement to meet the greenhouse emissions reductions levels (20–50 per cent) recommended by the scientific community (Stern, 2006). Even if the recommended reductions were immediately adopted as universal targets, climate change consequences would continue to escalate throughout coming decades. Climate change cannot be halted (Adger et al., 2007). The best that can be hoped for is that the rate of change might be slowed. Yet, political and economic leaders of contemporary societies continue to lack a unified political will in acknowledging and addressing these truths. The delays created by their stubbornness in clinging to 'next-time' decision-making patterns raises serious doubts about the extent to which the most disruptive impacts of changing climate might be alleviated (Cafaro, 2011).

In the 21st century, global climate change will require new political practices among states and other actors, including redistributions of authority. The Copenhagen Summit demonstrated just how difficult this will be if democratic and cooperative procedures remain integral components of the international community's response. To date, the differentiated interests of states have thwarted attempts to articulate widely accepted platforms for action. Breakthroughs that galvanise rapid and fundamental changes will remain rare unless individual states adopt leadership roles and unilaterally begin enacting responses. This is not to suggest that unilateral action by states can successfully address climate change, but rather genuine progress may well rely upon individual states adopting leadership roles to galvanise broader action.

Governments, economic actors and international policy-makers must accept increased levels of responsibility for developing and implementing mitigation and adaptation strategies. If they are to create and implement meaningful mitigation and adaptation strategies, they will be required to engage with global, regional and localised climate change consequences. Their responses will require them to differentiate their specific responses and spheres of accountability in dealing with the uneven impacts of global climate change. In turn, states and other political actors will establish a new range of legitimate activities as they construct new sites of political interaction (Edmondson, 2011).

New political institutions and sources of authority will be required to underpin the structures and administrations that emerge as the international political community grapples with the realities of increasing environmental security risks and eroded industrial prosperity (O'Brien and Leichenko, 2000). It seems likely that the volume and magnitude of these changes will not be accommodated within established democratic political processes. While the abandonment of democratic principles for governance is a deeply disturbing prospect, it is apparent that new political processes and structures will be essential for effective global climate change responses. At the very least, informed debate involving the global population takes time and this is a resource the scientific community believes we do not possess.

Climate change poses new and complex problems for a broad array of political and economic actors, ranging from those in high-order international activities to small-scale enterprises grappling with altered regulatory environments. Their challenges include dealing with changing production inputs and new market sensitivities, which are likely to become subject to greater unpredictability (Parry et al., 2007; Kjéllen, 2006, p. 8). However, these sites of economic and political activity also encompass sources of social and political vision. The changes they experience are likely to influence climate change mitigation and adaptation strategies, as well as changing economic policies within particular locations. The flow-on effect of global climate change upon political and economic actors will include revised values concerning the features and benefits of well-ordered societies (Cafaro, 2011). The sources and forms of authority that people create and shape as core political institutions are likely to be affected. The challenges climate change consequences present to states, intergovernmental organisations and other actors will alter their opportunities and capacities to assert themselves as authoritative agents.

These changing forms of social and political organisation and influence suggest that fear will play an expanded role in relations between states as less democratic forms of authority become important sources of social and political order across domestic and international communities. It will become increasingly important for political leadership to prioritise the survival of the planet over short-term economic growth. As climate change impacts disrupt existing patterns of habitation, production, trade and exchange, the political costs of attending to these challenges will diminish.

The international political community has behaved as though the borders between states are enduring and tangible sources of territorial delineation governed by the authorities representing particular peoples. The reality, however, is that these borders amount to superficial and recently imposed means of differentiating political entities. Political borders, most often, do not arise from the natural environment. Neither do they reflect levels of sensitivity or awareness among political communities to the ecosystems that criss-cross, intersect and overlap their physical environments (Vitousek et al., 1997). The privileged status of political borders reflects the reluctance of modern societies to accept their reliance upon the natural environment and its life sustaining resources.

Responsibility without borders?

Sovereign states have collective responsibilities for preserving international political institutions that enable the development and implementation of global climate change responses (Edmondson, 2011). Effective responses will only be achieved with appropriate attention to the distribution of relative costs and equitable burden sharing (Eakin et al., 2009; Roberts and Parks, 2007). The structures, institutions and authoritative status of international actors have been premised upon the capacities of states as sovereign entities. The contemporary international political community was formed on the basis of these perceived capacities and its key institutions have enabled their autonomous pursuit of industrial and economic progress. However, the political ramifications of global climate change require new understandings of the roles, rights and responsibilities that characterise states. These arise because the contemporary international political community has been formed on the basis of ideas concerning states as primary sites of political authority that no longer have as much utility.

Global climate change poses extensive threats to political order, social harmony and even the habitability of many locations (United Nations Development Programme, 2008). Its consequences straddle geological, ecological and ideological dimensions and impact upon all layers of the complex modern societies that characterise the contemporary world. While these disruptions are extensive and threaten the well-being of many communities, apocalyptic hyperbole is unlikely to be helpful in promoting the range of political changes necessary for developing mitigation and adaptation strategies. However, the experiences of protracted and less-than-effective international greenhouse emissions targets over the last decade suggest a business-as-usual approach to political negotiations and bargaining are unlikely to prove either timely or effective (Intergovernmental Panel on Climate Change, 2007a). As Hoffman (2002, p. ix) argues, '[w]hat threatens us is...an imbalance between the supreme legitimate authority' of states, and the 'feeble authority of collective institutions dealing with problems that transcend the states, or exceed their capacities.' At present, states struggle to translate their collective goals into international achievements. Even when they can agree on key goals and targets, they struggle to establish policies for managing climate change mitigation or to achieve collective strategies for redistributing technologies.

The new international political context created by global climate change raises the possibility that the expanded array of political actors and voices seeking influence might extend collective responsibilities among states (see Chapter 4). New sites of collective responsibilities, such as monitoring sea-level changes and limiting the production of ozone-depleting chlorofluorocarbons, might contribute to global strategies for addressing the multiple challenges of establishing new forms of economic endeavour and securing human well-being (Edmondson, 2011). However, empowering these new sites with political authority to secure vibrant and harmonious human communities into the future will demand considerable political leadership. Specifically, these changes will require resetting the balance between political and economic actors to privilege global commons principles.

Even at this early stage of witnessing an array of direct and indirect threats to human well-being arising from climate change consequences, it is apparent that these cannot be isolated to individual states. Climate change consequences are widespread, only partly predictable in their patterns of occurrence, and overwhelmingly disruptive. They cannot be isolated to groups of states or political coalitions or even to particular economic sectors across the world. While this observation might seem self-evident, it often remains implicit, at best serving a background role in international negotiations concerning global climate change.

States tend to prioritise their individual interests over others, including privileging these over collective strategies that might better preserve their futures. These dynamics point to some fundamental political challenges for the international political community and the forms of government authority that continue to hold legitimacy. They also pose new questions concerning the nature of relations between citizens and their governments, specifically in relation to the terms upon which states might be deemed to hold authority to act on behalf of their citizens. These political tensions are exacerbated by the ways in which states are accustomed to identifying and legitimating their national interests in terms of maximising their levels of economic growth (O'Brien and Leichenko, 2000).

The consequences of global climate change will unmake the successes of many economies. It will also unravel the political beliefs, assumptions and basic institutions that made previously shared visions of progress possible. Around the world, for at least the last couple of centuries, good government has been measured primarily in terms of the abilities to grow economies. Effective government has been measured in expanding production, increased profitability in trade and abilities to attract foreign investment. More recently, expanding the proportions of their populations engaged in skilled employment has been added to expectations of good government. These widely shared beliefs about the benefits of economic growth and dreams of expanding affluence have provided driving forces for development in the international political community, individual states and key transnational economic actors (Edmondson and Levy, 2008; Sprinz and Weiβ, 2001).

Responding to climate change consequences successfully will depend more upon the functional aspects of sovereign statehood and less upon states' shared legal status as central political actors (Biermann and Bauer, 2005; Postiglione, 2001). What states do, and the systems, mechanisms, agreements and international organisations they create to achieve their climate policy goals, will be more important than preserving historical legal notions of equality and authority. Accepting that states share common interests in self-preservation can lead to revised conceptions of authority. These will be important to the manner in which states identify, accept and act as their new collective responsibilities become more evident. By forming and maintaining international institutions, organisations and agencies to implement international agreements, states recognise their mutual rights and responsibilities to protect national interests. Such institutions reflect interdependencies and common interests among states and highlight the importance that states attribute to their privileged authoritative status (Hurrell, 2007).

Across the world in recent centuries, borders and security interests have been imposed upon the natural environment, carving a global ecosystem into discreetly named, owned and un-owned geographic regions. The challenge for states in becoming global guardians lies in seeing beyond the political divisions that human societies have imposed upon the natural environment (Arnold, 2011; Krakoff, 2011). However, as states collectively respond to climate change, seeking to establish and implement mitigation and adaptation strategies, across international, regional and local political spheres, they face new challenges. Global climate change consequences now require more universalised recognition of the limitations of human societies and their institutions in guarding and managing the complex systems that support life.

States and progress

Effective mitigation and adaptation strategies require political, economic and social actors to come to terms with the realities that even the best political will, most extensive financial resources, sophisticated technologies and creative problem-solving cannot enable human societies to exert control over their natural environment (Flannery, 2005). The physical world is comprised of complex ecosystems, temperature and gas exchange processes. These produce climatic systems and support all forms of life. Modern systems of production and patterns of consumption impact upon these physical processes, and, at least, some of these are deemed likely contributors to global climate change.

The best imaginable systems of economic and political organisation are unable to control the systems and processes that comprise the physical world (Flannery, 2005). The continued existence of human societies depends upon these uncontrollable systems, processes and feedback mechanisms. It is a basic reality that water, oxygen and fossil fuels cannot be manufactured. New technologies cannot remove the basic needs of human societies to access a diverse array of life-sustaining resources. Most importantly, the sovereign authority of states poses additional obstacles to the attainment of effective climate change responses. Artificially divided by political borders, the international political system relies upon the global ecosystem for resources that might support their quests for expanded profits, consolidated security or enhanced influential capacities.

The difficulties faced by governments and others engaged in responding to global climate change arise from established and persistent beliefs in the particular importance of humans as superior sentient creatures (Eckersley, 2004). These foundational elements of human political action and economic endeavour lie at the heart of the identities of human communities. Replacing these sources of social identity with alternative ethical structures is a significant challenge. It is additionally problematic because of the diverse range of values, ideologies and religious systems built upon these generalised beliefs. While modern societies are diverse, they largely share beliefs in their incontestable rights to utilise natural resources.

As a result, responding to global climate change does not simply entail generating effective policies and agreements. Instead, it necessitates alternative relations between individuals and their authorities; changed perceptions of the roles and structures of decision-making agents; and new sets of beliefs concerning interdependent ecosystems that do not unnecessarily privilege humans. Consequently, effective responses to global climate change will require some unpicking of the current fabric of the international political community (Young, 1994). Specifically, new conceptions of sovereignty, new notions of security and new ideas about what constitute appropriate practices within the international political economy will be required.

Maintaining an international community that is premised upon states as supreme holders of authority is problematic for responses to global climate change. Most especially, the manner in which states enjoy privileged status as legal and authoritative entities limits the emergence of alternative sources of authority that might more readily achieve effective climate change mitigation and adaptation strategies (see Chapter 4). One of the important challenges that arise from the privileged status of states as sources of political authority concerns the manner in which their interests are perceived and pursued. States currently enjoy a rights-based form of sovereignty that provides them with the political authority to collectively determine the values and direction of the international community. They have a right to determine what occurs within their own borders and to have their views heard and respected internationally. As a consequence, they are the pre-eminent features of the human-made global landscape and the natural environment tends to be ascribed value only as a source of economic opportunity.

The evolution of states in the 19th and 20th centuries established particular links between their sovereign capacities and their pursuit of economic progress (Edmondson and Levy, 2008; Philpott, 1997). In this manner, economic growth attained priority as a goal shared among sovereign states. However, these goals were independently pursued because of a general belief that an inherent and persistent security dilemma existed between states. In short, states perceive themselves to suffer from inescapable security risks arising from the presence and activities of other states (Buzan, 1993). These dynamics tend towards a political emphasis upon order as a fundamental organising principle and economic activities that provide states with benefits and advantages are given priority in their decision-making.

In generating international agreements, including those relating to global climate change, the international political community has, thus far, demonstrated a preference for adhering to broadly democratic processes. However, creating, implementing and maintaining climate change mitigation and adaptation strategies may lead states (and others demonstrating political leadership) to reconsider the forms of power they seek to exert as well as the nature and range of interests they claim to accommodate, protect and pursue. Historical models of sovereignty required states to apportion recognition to others to enable them to fulfil their functional capacities as sovereign authorities (Fowler and Bunck, 1995). Without the political legitimacy bestowed by sovereignty states lack the means to define their political interests and the legal right to protect them.

These factors are important in international climate change debates and policy-making because the interests experienced and pursued by states are inescapably linked with their domestic political goals. This adds complexity to debates concerning interdependencies among states and increases the difficulties of achieving a just distribution of burdens. Nonetheless, the norms and institutions that provide mechanisms for attaining order between states might also prove valuable in reconciling competing claims to rights. They may be able to utilise some common heritage principles that reflect states' environmental interdependencies to effectively demonstrate that protecting a shared ecosystem is a shared security priority.

States face significant challenges as they attempt to reconcile their national interests whilst responding to climate change consequences. Their roles as legitimate sites of political identity and international decision-making are being further complicated by the need to maintain authority and policy boundaries between the domestic and international spheres (Krause and Renwick, 1996). It is no simple task to think locally but act globally in an endeavour to serve the interests of both a national and international community. This is especially difficult for individual states whose authority is already contested, or who lack the political will or economic resources to meet the growing demands for domestic political action in support of collective common goals.

These new political dynamics are creating logistical and conceptual confusion among political actors because national and international interests have become impossibly blurred and interconnected. Previously, the ability to differentiate and clearly articulate national and international interests and obligations enabled international order. Now, however, the imperatives for states to provide meaningful protection of their territories and people, in order to fulfil their implicit promises of political stability, are being overtaken by imperatives to ensure planetary survival (Roberts and Parks, 2007). As atmospheric temperatures and the incidence of severe storms increase, and our abilities to feed and sustain a growing population are increasingly compromised by climate changes, it is impossible to divorce national interests from international interests. It is also impossible to divorce the political security interests of states from the ecological security interests of the international community (Mische, 1989). Attempts to privilege the political security of some societies over others by ignoring the diverse ecological ramifications of climate change are manifestly unethical.

Ethics, security and order

The processes of achieving effective international agreements can be streamlined by recognising that states hold particular moral obligations to adopt and enforce environmentally responsible practices because they are the basic organisational and territorial units of global politics (Low and Gleeson, 2001). Further, the political authority ascribed to states through the beliefs, practices and legal status of international sovereignty contains implicit ethical elements (Bonanate, 1995). These are apparent, among other things, in their pursuit and development of international agreements, treaties and conventions to manage or alleviate problems that lie beyond their individual capacities. For instance, sovereignty brings with it a set of expectations that requires states to protect their citizens. It drives states to promote prosperity by guaranteeing resource and territorial security. In this, prosperity–security cycle, states endeavour to maintain predictability in their relations with others.

Sovereignty carries additional ethical elements in terms of requiring states to maintain and protect their territories and to ensure their survival as political entities and authoritative structures through time. Within the international political community, sovereignty requires states to exercise their ethical predispositions by recognising the authority of other states. It is on the basis of this interdependent feedback cycle that states exert their claims to authority, achieve recognition and engage with others (Edmondson and Levy, 2008; Keohane and Nye, 2001). On the basis of these dynamics and political structures, states 'justify their actions' calling upon their responsibilities for 'security, prudence, safeguarding of the national interest and so on' (Bonanate, 1995, p. xii). In doing so, states and the international organisations and agencies they establish to act on their behalf, 'choose their behaviour from a variety of alternatives' taking deliberate choices even when these achieve very limited actions or interventions (Bonanate, 1995, p. xii). Most states, most of the time, believe that their decisions, ultimately, reflect the interests of their people, economies and governments. This hope is problematic in relation to global climate change because it creates obstacles to effective and timely initiatives by privileging decisions and interests that derive from expectations of rights rather than responsibilities.

The authority of sovereign states is fundamental to the responsibilities and obligations they assign to each other and underpins their capacities to fulfil international obligations. International agreements and institutions, including international law, derive from the existence of sovereign states, creating the basis of shared purposes and activities. Hence, basic international 'legal principles', such as 'mutual recognition', observing treaties and maintaining a 'functioning diplomatic system' reflect a generalised sense of 'common good' and an interest in international order (Hurrell, 2007).

As illustrated in Figure 11.1, international order, like the effects of global climate change, transcends the individual security and economic interests of states and concerns itself with the conditions that preserve peaceful coexistence. It requires states to collectively ensure security for their citizens, protecting them from violence, invasion, government by foreigners and the imposed interests of foreigners. When a

Figure 11.1 International ethics, reason and identity among states

group of states is conscious of these common interests and values, they are able to conceive of themselves as bound by a common set of rules and institutions (Bull, 2002, p. 13). These and other dynamics of international order enhance states' prospects of self-preservation. They encourage behaviour among states that alters their established identities and produce new orderly relations.

The complex interdependence that characterises relationships between contemporary states means that an international society of sorts exists (Luard, 1990, p. 3). The contemporary international political community clearly displays these features through the legal equality of states, their rights to fixed borders and reliance upon recognition for functional international membership. Human societies are centred on dense sets of relations across political, economic and social spheres, and it is these relationships that global climate change reorders. To date, however, the predominant shared goal between states has been order through the maintenance of their independent authoritative status, rather than concern for the planetary environment.

Order is a process constructed by states through their practised, habitual behaviours that regulate their conduct and establish expectations and routines that perpetuate these forms of behaviour (Luard, 1990, p. 62). Through these norms states create institutions to support further predictability and stability in their relations. They create 'rules ... that prescribe behaviour' and codify the accepted roles of others (Keohane cited in Reus-Smit, 1997, p. 557). This sustains a pattern of activity that constitutes international order. At the core of this international society are 'principled rules, institutions and values that govern' membership and behaviour (Finnemore, 1996, p. 18). Cooperative global action to address the causes and effects of climate change rests upon the existence of these conditions. It is membership of this political community that socialises states to construct and accept rules of practice. The major political challenge therefore lies in 'greening' these norms and institutions. As states recognise that they exist in reciprocal relationships, order creates a 'willingness, most of the time, to conform to the expectations of society' (Luard, 1990, p. 64). This provides opportunities to address global climate change by highlighting its impacts upon the orderly attainment of individual and shared interests.

A necessary element of an international society is the presence of order. As Dunne (2001a, p. 77) states, 'at a minimum, the survival of an international society requires a consensus on the basic principles of international order'. Order is a shared goal between states. An integral element of a 'society' is the presence of common goals, or 'shared interests' (Dunne, 2001a, p. 71). These common goals are formed because a society has to 'contain an element of common identity, a sense of we-ness' (Buzan, 1993, p. 332). Society, by definition, requires members who share values pertaining to the 'ends' that the members ought to attempt to achieve (Dunne, 2001b, p. 223). The most significant of these common goals has been order so that states may pursue their own development in relatively stable and peaceful political circumstances. This is an important dynamic that should be clearly expressed: international 'society' requires order, and norms and institutions are the mechanisms by which order can be achieved. Norms and institutions define international society and are the means by which states pursue commonly understood goals (Kratochwil, 1989, p. 11).

Rule structures in international society have particular importance because of their roles in creating order. The formal and informal rules that states and intergovernmental organisations create offer symbolic representations of their common values and beliefs, providing them with capacities to lead collective responses to shared challenges. Through their interactions, states are socialised to accept that these rules have been created by the international political community for a common purpose. States then accept that breaking the rules will bring condemnation from other members of the international society (Edmondson and Levy, 2008). Just as an individual obeys laws 'because he knows or considers, or feels, that he ought to', states also abide by norms and institutions because they recognise that their ongoing abilities to exert authority and contribute to order relies upon their acceptance (Manning, 1962, p. 114). Tying states' interests to the challenges of responding to global climate change provides genuine opportunities for new forms and expressions of order supported by cooperation.

By protecting international order, states are also protecting their rights by reinforcing rather than changing the established political structures. For this reason, strategies that effectively address climate change must coincide with normative and institutional activities beyond and beneath the boundaries of individual states. International order, pursued by states who seek the fulfilment of shared interests, remains a desirable outcome. These interests can be harnessed in support of collective goals in responding to global climate change impacts, and it is likely that doing so will galvanise mitigation and adaptation initiatives.

The prospects of international order in the 21st century require more than an old fashioned balance of power among states caught up in the pursuit of prosperity. Responding effectively to global climate change requires new levels of policy coordination and integrated bargaining mechanisms among states who accept that their authority includes collective responsibilities for global preservation (Young, 2002; Paterson, 2000; Haas et al., 1993). As these impacts escalate, global climate mitigation and adaptation strategies will increasingly rely upon effective leadership and institutions that enforce accountability. These will more readily be achieved if addressing climatic challenges are recognised as integral to strategies for maintaining order within the international political community.

Conclusion

In order to participate in international political negotiations and governance mechanisms, states must be able to ascertain priorities in their national interests and be capable of responding to the interests of others. While these dynamics among states and other political actors can be perceived to reflect conditions of enlightened self-interests, the existence and development of international agreements suggests that states are engaged in more than mere self-preservation (Halliday, 1994; Young, 1989). As global climate change consequences unfold, these political dynamics will become important features of international mitigation and adaptation strategies, especially in relation to managing the distribution of authoritative status as well as the relative costs of implementing agreements.

There are pressing imperatives for climate change mitigation and adaptation policies. Effective and durable policies will need to pay attention to the diverse rights, responsibilities and perceptions of the common good that exist among states and are shared within the international political community (Bonanate, 1995; Najam, 2005b, p. 245). However, physical interdependence among states does not of itself make it easier to achieve international agreement (Young, 2002). It is, therefore, time to engage with some of the international political dimensions of global climate change, including the manner in which it raises new questions regarding the nature and distribution of rights and responsibilities.

State sovereignty and international order requires social harmony among the diverse populations and identities that comprise the international political community. Ensuring the livelihood of their citizens and maintaining social and political order will require states to accept their collective obligations to support adaptation and mitigation policies that prioritise global well-being and the well-being of human societies into the future. Accepting economic challenges and relinquishing hopes of never ending economic growth and industrial expansion will be central factors for effective international climate change responses.

States are experiencing increasing pressure to take responsibility for climate change in order to justify the preservation of their sovereignty and authoritative capacities. States are responsible for domestic environmental practices and solving climate change issues in relation to their own territory, but they must also construct, apply and adhere to global climate change responses. If meaningful political authority is to continue to be exercised through states they will need to acknowledge the range of responsibilities that arise from their sovereign rights.

States are equipped with capacities for ethical action and decision-making. Most of the time, most states do not seek to invade or destroy their weaker neighbours, but rather form alliances with others they perceive to share at least some of their interests (Walzer, 2000; Barry, 1986; Krasner, 1982a). International agreements can provide valuable frameworks for interactions between international and domestic political actors and their interests if they are premised upon cooperation. By creating and maintaining complex international institutions and structures, states have collectively improved the international political system. Nonetheless, addressing global climate change is invariably difficult because it requires all states to recognise the superior significance of their common interests.

12
Global Guardians

Introduction

Popular notions of industrialisation and development are no longer viable and a growing global population cannot achieve affluent lifestyles configured upon Western images of progress (World Wildlife Fund, 2012, p. 6; Corson, 1994, pp. 206–207). Such aspirations inevitably collide with issues arising from global climate change and their political and economic implications. Across the world, governments, their citizens, international organisations, economic and social actors confront political challenges in resetting the ideas, values and practices that sustain their well-being. Cooperative international problem sharing will rely upon popularising new core values to ensure effective environmental management and sustainability. Whether or not human activities are drivers or mere contributing components of increased storms, redistributions in water and changed climatic patterns has become irrelevant. Also irrelevant are debates concerning potential rates and distributions of climate change impacts or arguments concerning the relative merits of various projection models. The world's ecosystems cannot absorb the impending resource demands of the currently expanding population: these physical limitations would have required major changes in industrial practices regardless of observed changes in climate.

These new political realities raise ugly questions about how the aspirations of the world's population should be ethically mediated to become ecologically sustainable (Markowitz and Shariff, 2012a; Gardiner, 2011). Intergenerational territory and resources security provide common themes to many of the ethics-related debates that ensue. Thus far, however, aspirations of ensuring that future generations enjoy resource security have proved weak drivers of international and domestic political and economic policies, and they remain noticeably absent from key international environmental governance mechanisms. These challenges are poised to intensify because the 20th-century goals of industrial development and political stability arising from the global trade of capital and carbon have been successfully promulgated as universal goals.

Historical visions of production and a broad social distribution of prosperity were configured upon the political values and images promoted by resource hungry, high energy consuming, middle-class Western societies. However, it is now clear the fragility of the earth's ecosystems, which have been heightened by global climate change and water redistributions, prevents its attainment (United Nations Development Programme, 2008; World Wildlife Fund, 2008). Degradation, extinctions and transformation have been the environmental effects of the rise of modern societies. Managing, accommodating and accepting these effects have long been integral to progress.

Industrialising societies have tolerated environmental damage as unwelcome but acceptable by-products of progress. Examples include air and water pollution, deforestation, loss of fertile soil through erosion, increased soil salinity and acid rainfall. Recognising the point at which these side effects become more direct threats has been a routine component of the rise and fall of local and national communities (Linden, 2006). In the 21st century, climate change provides the most recent lens through which to re-evaluate these accumulated impacts. Over successive centuries, human beings have pursued opportunities to enable more people to live more comfortable lives by utilising an ever-expanding range of natural resources and forming societies that have maximised the 'life benefits' of their collective activities. Such progress has necessitated environmental manipulation, resource use and exploitation (including over-exploitation) of those resources most highly valued as the sources of 'leaps in progress' at key points in history.

In all probability, reconfiguring how humanity conceives of its relationship with nature, and modifying modes of production accordingly, will be necessary to enable societies to adapt to the changes currently occurring in global ecosystems (Eckersley, 1992). A key question concerns where the tipping points lie in creating common survival interests because these will be essential to agreements about individual and collective rights to material goods, new expectations concerning opportunities and ideas of progress.

Perceptions of the environment as available for use in ways that prioritise human interests over the needs of other species have been founded upon deeply held convictions often supported by religious creationism or Darwinian science (Ponting, 1991, pp. 141–160). These perspectives conceive of humans as a superior species who are entitled to enjoy the bounties of the world's natural resources and ecosystems. As a result, the modern international community displays entrenched views of entitlements to resources that enable societies to flourish, increase their levels of prosperity and expand economic production. Questions of appropriate resource use have most often been evaluated in terms of their relative advantages for societies rather than their environmental impacts.

Religions, philosophies and values underpinning economic systems have each contributed and reinforced attitudes that have shaped the contemporary world (Burgess et al., 2003, p. 265; Ponting, 1991, pp. 141–160). Many of these have been premised upon inalienable rights to resources and diverse, life-supporting, environments. Through these various belief systems, human activities have been widely deemed to 'add value' to nature. These views have been widely apparent in modern states at least since the beginning of the industrial revolution and have informed many aspects of European imperialism, colonisation practices and post-colonial political enterprises. They have been vital in creating the political and economic systems and activities of states that have led an international pursuit of political and economic progress through industrial production, and also those who have been less successful in achieving economic growth. Within dominant Western societies such values have produced systems of property ownership and industrial practices grounded in human-centred ownership of the natural environment and its resources, promoting property rights as central features of good societies.

Responding to these challenges necessitates attention to ethical and other values-based decisions and judgement. For the international political community, these considerations include deciding who holds the right to prescribe patterns of development and progress? Who holds the right to deny opportunities of prosperity? Who might exploit resources in building the international political economy of the 21st century? How do we begin to believe that progress does not bring longer secure lives for more people supported by a greater range of material goods? It does not matter how lean and clean societies become because there are so many more people to support, and so many more of these are becoming users of cars, light bulbs and packaged foods (Linden, 2006).

Into the 21st century revived conceptions of a global commons potentially support an evolution in the nature of statehood as issues of impending scarcity overshadow property rights issues among states (Dryzek and Schlossberg, 2005; Eckersley, 2004; Paterson, 2000). In appreciation of climate change, some states are demonstrating their willingness to consider key resources in a broader context of cooperation. Admittedly, to date much of this has occurred when states have been forced to recognise that their own choices rely upon the decisions of others, such as in managing the problems of transboundary pollution (Edmondson, 2009; Young, 2002). These expose tensions between the political and ideological requirements for continued economic development and the capacity of the natural environment to sustain established notions of progress.

One of the political impacts of 21st-century global climate change is that a new environmental/ecological conception of a global commons is emerging as governments increasingly appreciate that rainfall, water currents and nutrient flows within water sources underpin human life. As a consequence, the international political community may develop a view of the global commons as requiring special protection. Such a development, however, would require a broad and balanced appreciation of the extent to which such commons may be sacrificed to ensure orderly and equitable global development. These might include damaging an array of ecosystems within particular states, while for others it may involve accepting the inevitability of the extinctions of certain species, or foregoing certain forms of industry.

An additional political dimension lies in the historical backdrop wherein discrediting prophets of doom has long been part of the pursuit of progress (United Nations Development Programme, 2008). Nonetheless, the ranks of climate change believers have swelled to include the majority of the relevant scientific communities supported by increasingly vast administrative bureaucracies intent upon the systematic dissemination of their message. While there continue to be influential climate change sceptics, and others who contest the physical dimensions of global climate change, these are substantially outweighed by those who conclude that climate change is occurring at rates more rapid than even the direst projections of a decade ago.

Entrenched views, beliefs and faith in the 'progressive' abilities of humans reflect dual perceptions of superiority over the natural environment and the development of creative solutions through new technologies or breakthrough ideas (Weisman, 2004). Many who argue against 'prophets of doom' value humans and the progress of societies above the preservation of other species or the integrity of any ecosystem. They continue to expect that humanity's best and brightest will master the challenges presented by 'nature'. The presence of these views within the international community has contributed to the slow responses of key political actors and their low levels of willingness to exercise morally founded leadership.

Enduring beliefs in progress, and its many cycles of progressive uncertainty – endeavour – breakthrough have created modern industrialised societies and governance mechanisms, with corresponding expectations that 'we always find timely solutions'. To some extent, there is good reason underpinning hopes that new technologies might provide solutions that currently only a few might imagine. For instance, in the 1960s as the world's population dramatically increased, new technological breakthroughs produced substantial leaps in agricultural production by developing disease resistant and dry tolerant rice and wheat varieties (Deb et al., 1983). In a similar manner, in the early 21st century, concerns over depleted fossil fuel supplies pressed states into exploration of new potential energy resources. Consequently, India, China and Japan have pursued lunar explorations in search of minerals resources that might support their production activities and lifestyles (BBC News, 22 October 2008).

Underpinning the modern international community is a collective belief that progress-based solutions will always emerge from new technologies, new practices, new resources and new efficiencies (Stern, 2009; Weisman, 2004). These optimistic attitudes account for many of the successes achieved throughout history, but they also account for the snail-paced responses of states towards the challenges of climate change. Refusing to acknowledge that new and changed conceptions of progress, industrialisation and forms of societies may be necessary leads to low levels of individual willingness to proactively respond to climate change. While the worst effects of climate change continue to be forecast beyond the lifecycle of the majority of present-day individuals, there is a propensity to believe solutions will be found that validate current attitudes and practices.

Among the issues at stake are dominant notions regarding the tolerable and equitable distributions of responsibilities, challenges of development, benefits of progress and appropriate relationships with the environment. These complex political problems are not limited to ethics and justice for a global commons. The political and security tensions presented by global climate change demand new solutions that include establishing new relationships between people and their environment as well as between states, corporations, citizens and civil society. As has been noted: '[c]onsumption practices are fundamentally political' (Burgess et al., 2003, p. 268). Consequently, effective climate change responses must begin with recognition that: '[l]arge organisations, public authorities, the military, and commercial companies, are largely responsible for the bulk of human-environment transactions which lead to damage to bio-physical systems' (Burgess et al., 2003, p. 268). Although these issues are not new, climate change has brought them into sharp relief.

Debates of this nature do not privilege the environment over human communities but rather seek sustainable practices of development that permit an increasing global population to continue to progress in universally desirable ways. In so far as there is an ethical element, this concerns our relationships with future generations and with nature. As the pursuit of progress depletes the earth's carrying capacity and intensifies climate change consequences, we are forced to consider what might be left for future generations (World Wildlife Fund, 2012; United Nations Development Programme, 2008; Flannery, 2005; Burgess et al., 2003). How the international community responds to global climate change and maintains orderly societies will depend upon the values we apportion to nature and the resources provided by the natural environment (Gardiner and Hartzell-Nichols, 2012; Edmondson, 2011).

Whether we accept arguments that societies and individuals are ethically obligated not to exhaust or fundamentally damage the earth's environment, such that subsequent generations are faced with an uncertain and less secure future, will depend upon our views of rights, responsibility, authority, freedom and security. International responses to global climate change will be directly shaped by how we rank our own comfort alongside the needs of others, including our grandchildren. Studies from around the world have shown that although fear and anxiety arising from environmental changes are widely understood and communicated, they have had little impact upon the daily behaviours of individuals (Burgess et al., 2003, p. 271).

Commons versus order

Recently, as states have grappled with an increased array of social, political and economic challenges arising from the actions and production methods of other states, they have also become more aware of their inherent interdependencies. This awareness has resulted in unprecedented numbers of international agreements and organisations to match solutions to identified problems (Edmondson and Levy, 2008). States have repeatedly attempted to balance and assess an array of moral factors including how to manage the costs of climate change impacts. Should industrialised states recognise their contributions to an impending energy crisis that will further prevent economic growth in developing states? Should developing states, with high rates of population growth and burgeoning industrialisation, enact regulatory policies to encourage greater environmental awareness and responsibility? For the international community, the most significant question then becomes to what extent and under what circumstances can it impose sanctions, restrictions and inducements to moderate the conduct of individual members? Viable answers remain problematic because such interventions fundamentally conflict with the international legal rights of statehood that enshrine sovereign independence and freedom from external interference.

Although sovereign authority and legal equality among states have been central features of the modern international community, states also convey ethical decisions through the responsibilities they create for themselves and others. The widespread adoption of sovereign statehood as the legitimate source of political authority encourages states to adopt and activate certain ethical decisions (Edmondson and Levy, 2008; Bonanate, 1995). Just as sovereignty provides states with responsibilities for territorial security, it also compels them to seek the well-being of their citizens. These functions of sovereignty require more than freedom from invasion, extending to a need for social harmony and secure access to life-sustaining resources. Hence, states are expected to ensure various means of livelihood for citizens, and to support the maintenance of social and political order (Walzer, 2000; Krasner, 1999; Friedman and Starr, 1997; Bonanate, 1995).

States are also engaged in exercising ethical decision-making when they acknowledge the rights of other states to make and implement laws, establish law enforcement agencies and other government agencies that enable them to function as sovereign authorities (Krasner, 1999; Bonanate, 1995). They exercise ethical decision-making when they defend their territorial boundaries and undertake other activities to protect their citizens (Barnett, 2004; Bonanate, 1995; Barry, 1986). By repeatedly exercising judgment, developing and pursuing policies of various kinds, and interacting with each other, states utilise what Nardin (2008) calls the 'moral element' in international law. Ethical decision-making is apparent when states exercise 'rights of independence, legal equality, and self-defence', in conjunction with 'duties to observe treaties, to respect human rights, and to co-operate in the peaceful settlement of disputes' (Nardin, 1983, p. 233). This capacity of sovereign states is an important source of hope for the international community as it seeks to identify solutions to global climate change impacts and future opportunities for sustainable human communities (Edmondson, 2011; Amstutz, 2005).

Becoming guardians

Treaties, conventions and international organisations highlight the abilities of sovereign states to develop new forms of international order to manage and preserve global commons resources. Treaties that address climate change, ocean resources and greenhouse emissions, along with other conventions and international agreements, reveal the acceptance by states of underlying moral decision-making. The principle of sovereignty preserves states as primary international political entities with moral rights and responsibilities and independent authority over people and territory. Developing a sustainable future vision requires states and the international community to reveal fuller abilities in employing common heritage principles and to develop international law that supports their ethical decisions (Markowitz and Shariff, 2012b; Camilleri et al., 2000; Paterson, 2000; Haas et al., 1993).

Morality and ethics among states

There is an extent to which discussions of ethics in the international community suffer from a false dichotomy whereby decisions and actions are deemed 'moral' or 'immoral'. This produces confusion by blurring political factors and interests, with economic and ideological factors and values. Climate change mitigation and alleviation strategies rely upon international law and supporting domestic legislation, regulations and multilateral commitment to attain their goals and targets (Vogler, 2005; Sprinz and Weiß, 2001; Oberthür and Ott, 1999; Rittberger, 1993). In the 1990s, negotiations emphasised fairness and the international community attempted to develop institutions to manage climate change consequences. Underlying ethical considerations concerned the distribution of burdens and long-term impacts upon states' abilities to achieve their national interests (Edmondson, 2001).

In efforts to manage the impacts of climate change, states and the intergovernmental organisations they created have confronted multiple, complex moments requiring ethical decision-making. These have resulted in protracted periods of negotiation and bargaining, target setting and revision, prior to agreement implementation and full institutional development. As a result, international agreements concerning the management of climate change problems have often involved protracted periods of negotiation and discussion, even when they have not solved problems or taken positive steps towards mitigation and adaptation (Edmondson, 2011). The protracted negotiations and persistent blockages to meaningful multilateral targets in the Kyoto Protocol are merely examples of these dynamics.

Changing our understandings and expectations of states as sources of political identity based upon collectively approved modes of international behaviour, may transform the relationship between their ethical decisions and functional status. Rather than the ethical decisions of states arising from, and being shaped by, their functional status as independent authorities, it is plausible to imagine a dramatic transformation in this relationship. The scale of the climate change challenge, coupled with the range of new entities gaining voices in international political discussions, may reconfigure the states' moral status as well as changing their functional capacities (as discussed in Chapter 8). The capacities of states to find collaborators and partners to address global climate change will depend upon an ethical relationship with the environment – past, present and future.

Ethical political behaviour is becoming more pertinent as global climate change consequences replace rights-based views of states with notions of responsibility. This change is occurring in conjunction with a shift away from rights-based views of states and production practices in favour of 'a sense of obligation' (Bonanate, 1995). As the international community diminishes the rights of states to autonomy, by injecting moral considerations into global climate change responses, states will also use their authority to construct mutually constraining codes of behaviour. In this way, they are able to protect their territories, citizens and authoritative status. In the face of global climate change, the most potent national interest is self-preservation with ethical behaviour becoming a more substantial component in common perceptions of national interests. Injecting ethics into national interests assists in shifting states' concerns from inward-looking self-interest to outward-looking concern for themselves and others.

It is important for states and other international actors to develop common goals and 'moral' interests to enable them to respond to global climate change (Hurrell, 1995). Recognising their responsibilities for managing a global commons will be essential for their abilities to respond as global guardians. In climate change responses, states are simultaneously advocates of their citizens and sources of the rules of their coexistence. They hold responsibilities for managing uncertainty and threats to the well-being of their people. Consequently, in the future it is likely that states will seek to create complex international agreements, treaties and practices that extend their ethical decisions and achieve their collective goals. The consequences of global climate change and related mass migrations of people, changed structures of order and new modes of sustainable production will constrain other policy options.

By establishing international institutions, states equip themselves with agencies and structures through which their national interests are enacted (Krasner, 1982b). Precedents in the areas of human rights laws and conventions suggest future generations might observe that universal responsibilities among states did not emerge as a result of global climate change, but that new forms of behaviour developed on the basis of earlier commonalities (Abi-Saab, 1997). However, a capacity to act morally does not mean that states will do so. Neither does knowing what a moral action might entail necessarily mean that states will act accordingly (Barry, 1986). Acting immorally does not reduce the ethical decisions (what some might call the morality) of states because their responsibilities and capacities to choose between alternative actions remain unaltered (Bonanate, 1995). However, the presence of international regimes, and their contingent institutions, increases the likelihood of both.

States rely upon international institutions and structures to create international law and organisations that reflect their perceptions of common responsibilities (Barnett, 2004; Bonanate, 1995). As Bonanate (1995, p. xi) observes, although states 'justify their actions invariably and exclusively in terms of security, prudence, safeguarding of the national interest and so on, they always choose their behaviour from a variety of alternatives'. Their decisions always reflect their evaluations of rights and responsibilities as well as risk. Thus the international community routinely recognises the 'rights and duties' of states, and more than lip-service is given to limiting their abilities to inflict harm against one another (Doppelt, 1978).

International institutions support the communal goals of the international community. They do this in much the same way that constitutions provide a locus for the political values and world views of individual states by providing their institutional framework and equipping them with jurisdictional capacities. Within this institutionalised decision-making environment, states exercise their jurisdictional capacities, rights and responsibilities. These institutions support states' abilities to identify policy-making and implementation priorities (Rittberger, 1993). This dynamic holds particular importance when states attempt to set new communal priorities, such as in addressing global climate change. Over time, international institutions will assist in establishing acceptable outcomes and cost–benefit–authority distributions.

Conclusion

Fabricated notions of environmental exploiters and guardians create a false dichotomy that sustains particular patterns of progress and development (Krakoff, 2011). On the one side are the willing supporters of practices that exploit natural resources to sustain, promote and advance particular desirable lifestyles. It is easy to identify them as industrialists, developers, consumers and politicians who promote constant improvements to standards of living and the artefacts of progress as human rights. It seems less than fair to include among them the unwitting everyday consumers who support industrial and political practices chosen by their governments and economic leaders through routine lifestyle choices, including the parents who seek to provide more for their children. From a wide enough lens, it is possible to incriminate the vast majority who either experience, or seek to experience, the benefits of a modern economy.

On the other side of this dichotomy are those who advocate a more responsible, informed and less destructive association between human societies and the environment. Their explicit claim is that humanity has an obligation to protect the natural world and that more should be done to halt animal and plant extinctions and to stabilise vulnerable ecosystems. They lament the environmental damage arising from deliberate and unwitting ecological interventions and agitate for social, political and economic measures that will either 'freeze' or 'reverse' these undesirable impacts. Their goal is to find some balance in human activity that will be environmentally benign such that the planet may be stabilised according to some highly principled and contextualised notion of what is natural.

Evident in both of these positions, however, is a notion of humanity's relationship with the natural world that has underpinned progress and its effects to date. Both environmental exploiters and guardians perceive of themselves as superior to, and separate from, the earth's ecosystem and thus have a natural right to manage it as they see fit. Their positions represent a false dichotomy insofar as neither recognises the embedded position of humanity within nature. In both cases, humans and their political associations have deemed themselves the legitimate controllers of nature and share a common satisfaction from exercising human dominance over the natural environment. The real debate should not be contested over who has caused climate change, or who should respond to its impacts, and how. Rather, it should focus upon the ecological impacts of preferred lifestyles and how climate change will alter the ways in which everyday people live, how societies function and what economies produce. The fundamental challenge then becomes how to better manage the aspirations and relationships of all people in an ethical, equitable and sustainable manner (Arnold, 2011).

It is necessary for the political, economic and social structures that have been humanly constructed to re-evaluate and reform their associations with the natural environment. Failing to do so will result in greater competition, costs and conflict between societies in the pursuit of increasingly scarce and expensive resources. The climatic consequences of an unregulated pursuit of progress have accumulated, producing environmental consequences that are impacting upon the activities of states and societies (Cafaro, 2011). These are currently experienced as a steady decline in the earth's ecosystem to sustain a universal pursuit of material prosperity.

The futility of debates and approaches premised upon how climate change might be 'fixed' or 'negotiated' are yet to be realised. Faith in human ingenuity and the mythology of progress, that there is no challenge that cannot be overcome with enough effort and resources, reaffirms our superiority over the natural world. Nonetheless, the planet is changing and nothing can freeze it in time. Currently, dominant notions of progress are on the cusp of being recognised as at an end. They have been premised, throughout history, upon the subjugation of the natural environment. Growing popular appreciation that the earth is a closed ecological system, in which resources are finite and by-products accumulate, herald awareness of a genuine need to acknowledge the effects of human practices (Weisman, 2004, pp. 112–128). The environment may no longer be considered as separate from the foundation of human society, a canvas upon which progress can be recorded with impunity.

This undertaking will need to be led by political, economic and social leaders informed by an appreciation of the past and credible visions for the future. It will require courage, conviction and fortitude to see beyond national interests, electoral and economic cycles and the veil of inherited ideologies. However, leadership without broad-based support is futile and herein lies a further source of debates for the 21st century. Accommodating the effects of climate change will require a universally supported commitment to lifestyles and forms of progress that are less ecologically intrusive and more ethically distributed. As observed by Burgess et al. (2003, p. 285), ' [i]t is asking too much of the consumer to adopt a green lifestyle unless there is a social context which gives green consumption greater meaning'.

Ultimately, leadership only plans a course of action and it falls upon societies to become implementers. Which states and societies should accept responsibility for creating notions of progress that are less environmentally destructive are smokescreens to conceal agendas of national self-interest. Mature societies need to acknowledge a more fundamental debate about how to universally promote desirable and ethically distributed patterns of progress. Thus a fundamental question for the 21st century is: Can diverse, competitive and politically independent societies create the leadership and communal will to reform their relationships with the natural world without sacrificing the aspirations of any of their peers?

Disturbingly, the politics of envy and avarice still appear more compelling than any appreciation of an environmentally informed impetus for reform and the prospects for change do not appear promising. Ideologies that satisfy unreflective and vested interests, creationist and Darwinian notions of innate human superiority (Ponting, 1991), and a perversely non-progressive view of human nature (Huntington, 1993), suggest such fundamental reform is unlikely to occur quickly enough to maintain the earth's ecology as recently experienced. Consequently, it may then become necessary to confront uglier decisions about which people's dreams of 'the good life' will be perpetually denied. After all, 'normative (moral) values pervade contemporary discourses about consumption levels and practices, especially in the context of equity issues between rich and poor, and these can sit uneasily alongside scientific discourses which seek to maintain a value-neutral position' (Burgess et al., 2003, p. 265). The moral challenge then becomes the extent to which the global commons may continue to be sacrificed in order to maintain inequitable patterns of progress and development.

Conclusion: Why Global Responses Take Time

This book took far longer to complete than we initially anticipated, for many of the same reasons that have stymied rapid global responses to climate change. Life and other urgent concerns, the changing science, cycles of optimism and pessimism and continued scepticism all contributed. The simple question that gave rise to this project was: If addressing global climate change is a moral imperative, why are international responses so difficult to achieve? Our research confirms the assumption underpinning this question: international responses to global climate change are difficult to achieve, and the immensity of the problems create political obstacles that require political will and determination to resolve. Responding effectively to global climate change requires the international political community to collaboratively find solutions, and this is especially difficult given the magnitude of the problem and the diversity and number of necessary participants. Addressing climate change is so vexed because its consequences extend into a great many fundamental aspects of modern lifestyles – both lived and imagined. These same lifestyles are posited as potential or significant contributors to environmental problems. Accommodating so many uneven impacts and fundamental imperatives leads to responses marked by fits and starts, ambiguity, contested claims, conflicting interests and, ultimately, fatigue and malaise.

Nonetheless, three broad trends are apparent in ongoing debates about how best to respond to the global climate change conundrum. First, global climate change consequences will continue to affect human populations and their societies for decades to come with escalating impacts upon habitats and livelihoods. Second, conceptions of the world as divided along East–West or North–South axes that were relevant to understanding global events in the 20th century are now of limited utility and new political tensions demand resolution. Third, human populations might well relocate themselves and develop new forms of production in order to live in a climatically changing world. These changing social and environmental conditions will create levels of unpredictability in international political affairs, global production, investment and trade markets. These observations confirm global climate change as a threat to future aspirations for prosperity, the democratic formation of abatement strategies and the orderly conduct of relations among climate change impacted states.

Since the middle of the 20th century, there has been a steady shift away from the principle that state sovereignty, based upon the primacy of state-based interests, should be protected without regard to its consequences for individuals, groups and organisations.[1] New political contexts created by global climate change require the international community to cultivate viable political responses to the many effects of climate change, including the problems of maintaining current expectations about desirable expressions of order, prosperity and democracy. Viable responses will require political actors to discard conceptions of security and insecurity that do not take account of these changed political circumstances. This is proving extraordinarily difficult because there are so many different interests and agendas to reconcile. For instance, many scientific research projects rely upon funding sources that are far from neutral, and many advocates of new technologies hold political and economic interests in their adoption. It is not surprising that in seeking responses to global climate change, people and organisations pursue strategies that maximise their benefits or advantage their interests.

States, and their leaders, are understandably reluctant to establish and pursue greenhouse gas emissions targets that increase their vulnerability within the international political community. For states, their continuing quest for economic prosperity through international trade relations is part of the broader context for reducing their greenhouse gas emissions. Most economies are now partly or even largely reliant

---

[1] Contributing to this has been the international community's acceptance of the principle of collective security and the democratising project of Western states which ultimately produced an array of human rights instruments. In both regional and international politics, sovereignty is now taken to imply respect for an extensive range of human, resource and labour rights. These have created new ordering principles in political affairs that potentially limit state power as an essential component of sovereign authority.

upon their abilities to sell the commodities they produce to markets in distant locations. Many states and major economic actors fear the possibility that at least in the short term, reducing greenhouse emissions might mean accepting reduced profits, smaller market share, expensive changes to their industrial practices, electoral unpopularity, changed status within the international political economy and lower levels of economic growth.

Across the world, the spectre of economic regression evokes insecurity among political actors and their constituencies, and global awareness of these vulnerabilities has increased in the 21st century following the global financial crisis. Awareness of economic vulnerabilities tends to delay decision-making in responding to global climate change. Avoiding future risks seem less urgent political imperatives when apparently immediate risks of financial catastrophe dominate the political landscape. Even during periods of economic and political stability, governments and other economic actors seek predictability and certainty when they make policy decisions, and such certainty is not possible in relation to the effects of global climate change.

As global climate change implications become more pressing, efforts to manage and regulate the uses of the natural environment, the resources it produces and the forms of production it enables must begin with the coordinated governance of a great many facets of human activity. As observed by de Coninck, Fischer, Newell and Ueno (2007, p. 351), '[g]iven the broad span of mitigation and adaptation options, efforts on the climate front obviously will overlap with those in the areas of energy, air pollution, biodiversity, agriculture, development, and public health'. Currently, governments individually recognise that uncoordinated and unilateral decisions and actions will provide insufficient protection from climate change risks but remain incapable of moving beyond 'wait and see what others do' approaches. Nevertheless, we remain hopeful that effective mitigation and adaptation strategies will gradually emerge because governments and political leaders have previously acted in accordance with the precautionary principle embodied in the UN Framework Convention on Climate Change. In their combined efforts to hurry slowly, they have achieved political progress in setting and enforcing meaningful targets. In coming years, the balance between timeliness and political acceptance will shift as many experience the 'predictable surprises' of a changing climate. New political imperatives that favour timeliness over an exhaustive quest for further knowledge will likely cause many actors to shift their attitudes away from perceiving that 'doing nothing...is the best alternative' (Science and Environmental Health Network, 2000).

The politically, economically and socially integrated nature of the international community offers a wide variety of climate change mitigation policy options that could commence at the national level and be extended beyond national borders. Internationally these reforms run up against the structural nature of the global economy. Achieving international reform in this area is difficult because of the threat posed to competitive national advantages in the market place. However, the international political economy is also replete with instances of economic integration and cooperation when actors identify shared interests. We agree with Stern (2009, p. 8) who observes that:

> ignoring climate change would result in an increasingly hostile environment for development and poverty reduction ... [and trying] to deal with climate change by shackling growth and development would damage, probably fatally, the cooperation between developed and developing countries that is vital to success. Developing countries cannot 'put development on hold' while they reduce emissions and change technologies.

The political challenge for all states lies in finding new ways to secure their economic well-being and aspirations, and this provides potentially common ground for collaborative problem-solving.

Ultimately, the myriad consequences of global climate change presents people and governments with new questions concerning how societies might be organised and sustained into the future. Aspirations to maintain '[e]conomic growth and development' rely upon the earth's capacity to meet 'increased demand for natural resources in general and for energy resources in particular. Reorienting this growth trajectory requires investment in building new infrastructure, new capacities, and new institutions', as well as new ideas about political security and new relations between people and governments (United Nations Committee for Development Policy, 2009, p. 9). For these reasons, viable solutions to the many challenges of global climate change lie within the realm of politics rather than economics or science. While new forms of economic association will contribute to climate change responses, and new scientific knowledge and technologies will also be important, the impetus for change will come from the political sphere.

Political commitment has always been an underpinning component of initiatives and developments to transform societies. Confronting and addressing global climate change will be no different as '[d]ecision-making about the appropriate level of global mitigation over time' must also include risk-management processes designed to achieve sustainable mitigation and adaptation to global climate change impacts (Barker et al., 2007, p. 27). Such steps will only be effective if political leadership ensures decision-makers are able to take simultaneous account of the short- and long-term economic and social costs and consequences of climate change. While mitigation policies and strategies are many and varied, and each faces idiosyncratic barriers and incentives, politics provides both the vehicle and means of reconciliation. Politics, more than economics and science, is able to unify, coordinate, mobilise and empower individuals to adopt the necessary values, behaviours and aspirations to ensure desirable outcomes.

States' mutual interdependencies are expanded by the political challenges of developing international climate change mitigation and adaptation strategies. Recognising and responding to these changes in their political context, and their shared spheres of authority and governmental capacities, represents an additional step in their evolution. 'The bottom line is that designing international coordination mechanisms that work requires that countries agree on the goals, share similar priorities, and are willing to allow domestic policies and programs to be influenced by international imperatives' (Lee, 2009, p. 71). Recent crises within the European Union are illustrative of how difficult these can be to establish and maintain. It is for this reason that politics is the key to addressing climate change rather than technology.

Reconfiguring political and economic systems to ensure sustainable and prosperous futures for the world's people and the global community will be crucial as climate change impacts unfold. International political solutions that achieve greater equity in the distribution of economic prosperity and well-being will need to include effective technology transfers between states. These are unlikely to occur on a grand scale through philanthropy because they potentially reduce the profitability of some industries, and diminish market competitiveness for some states and economic organisations. Nevertheless, the 'private sector will play the major role in financing the additional investment required, but support must also be provided...through international cooperation' (United Nations Committee for Development Policy, 2009, p. v).

Achieving timely responses to climate change sit in political tension with democratic processes. The only genuine way forward is for the development of political will and unity that requires political

Conclusion: Why Global Responses Take Time   223

leaders who accept responsibility for creating new political visions and expectations among their peers and constituencies. Considerable grassroots support has already been cultivated, but the political leadership required to capitalise on this has, to date, been erratic. While many governments have the legislative and programmatic capacities to support the introduction and adoption of strategies to address activities that contribute to (if not cause) climate change and to respond to their effects, the political costs of doing so threaten to be electorally overwhelming. The realities of electoral popularity in democratic societies are not diminished by climate change. Consequently, initiatives have been modest because electorally responsive governments favour short-term political expediency and economic stability.

International agreements, such as carbon taxes and carbon trading schemes, are likely to be significant for establishing new economic and industrial practices and systems. Over time, they may prove as socially transformative of daily human life as the invention of electricity and the motor vehicle. However, carbon trading schemes or taxes, increased water management systems and less environmentally intrusive food and energy production processes ultimately challenge the established political visions of developed states. They call into question practices that are fundamental to the long-term security of all people living in all states. As industrially developed and developing states struggle to reshape their economies and respond to global climate change, many will cling to outmoded visions of progress and prosperity in spite of increasing threats to their well-being.

Energetic political leadership that supports revised visions of progress and prosperity will be needed to secure economic well-being and territorial integrity as states face unprecedented climate and energy challenges. Political leadership will be required to manage altered relationships between political and economic actors and to establish new political structures to support internationally negotiated greenhouse gas emissions targets. Establishing new modes of behaviour among states will form part of the package of new 'rules' within the international political community that will emerge alongside revised expectations of the distributions of rights and responsibilities among political and economic actors. In turn, these new rules will require states to cultivate, support and enforce new modes of behaviour among their citizens. These might be expected to include changed visions of prosperity and progressive political arrangements that better reflect and protect their resource interdependencies. Altered political and economic imperatives associated with securing energy supplies and supporting growing global populations will be important to the policies, practices and institutions that emerge to support 21st-century social and political associations.

While crystal-ball gazing may be notoriously unreliable and ineffective as a source of policy-making, and computer-generated models remain contentious, both have proven useful in establishing the key social and political visions that have created the contemporary world. However, it is timely to revisit ideas about the nature of prosperity and progress, and the structures that support them, to develop effective global climate change mitigation and adaptation strategies. Overarching international political arrangements and governance mechanisms, such as intergovernmental agreements and agencies will be important in supporting these changes. These new political realities are as inevitable as increasing atmospheric temperatures and changed rainfall patterns. Some of the international security and peace-management mechanisms, created in response to experiences of international conflict during the 20th century, might also be useful in reducing the risks of global conflict arising from climate changes.

In the 21st century, new energy pressures will increase threats to the security of people and governments, challenging the abilities of existing forms of political authority and governmental structures to provide political stability and economic prosperity (Anadon and Holdren, 2009; Yohe et al., 2004; Barnett, 2001). As climate change impacts alter patterns of food and energy production and consumption, change water distributions, alter seasonal climatic patterns and change habitability across the world, they will also change the nature of relationships between states and their people. Relations between states will also be changed, as will their established practices of authoritative recognition and abilities to treat each other as legal and political equals. In the past, episodic changes or evolutions in internationally recognised forms of political authority, arose from changing expectations among people alongside new ideas of social and political organisation and new economic practices (Philpott, 1997). There is every reason to expect that global climate change constitutes a trigger for a further evolution in international norms and politics.

The last major evolution in sovereign states coincided with the development of ideas concerning new forms of collective responsibilities in international security and peacekeeping arrangements. These are likely to provide a foundation for further political transformation as global climate change impacts alter current configurations of political power. In the contemporary world, states are already embroiled in myriad complex political interdependencies linked with their trade and security arrangements, many of which occur through collective actions and international agreements. These are likely to increase as global climate change consequences reveal new geophysical interdependencies.

Alongside collective and cooperative undertakings, states have also endeavoured to consolidate their rights as independent political entities. Through global climate change, the 'shadow of the future' has shortened to reveal the natural environment as an ongoing site of contested authority between states (Bearce et al., 2009; Paterson, 2000). This is more than a mere backdrop to global climate change: it is a core political reality. It will take time, political will and acceptance of new ideas to support political change based upon international and collective responsibilities and new aspirations of human security.

We are convinced that there are two clear reasons for optimism. The first of these is that ideas have previously, and recently, dramatically reconfigured the international political environment and transformed the capacities and identities of states (Philpott, 1997, pp. 15–47). Ideas were responsible for an evolution in what modern states should look like, how they should be constituted, their capacities for internal and external legitimacy and their recognition of each other. The transformative effects of ideas are rarely immediate. Their impacts take time to percolate through political and social spheres and processes, acquiring adherents who support the goals, aspirations and prescriptions of new ideas. Over time, these people and political actors become important sources of pressure upon a range of policy-making levers and individuals. In all political and social evolutions, there are setbacks, false starts and inappropriate choices. If, however, the ideas are robust and become widely accepted then changes ultimately take effect.

Over a roughly 65-year period, from the beginning of the 20th century, modern sovereign states emerged based upon the idea that national identity should be the basis of sovereign authority. At the time, the effects of this transformation were barely imaginable. Addressing the causes and effects of climate change will result in an even more significant transformation. Awareness of climate change, and the need for political action to manage and ameliorate its consequences, will not fade away because there is a widespread groundswell of demands for clearer and more effective responses. While internationally orchestrated and effective responses to global climate change have not yet been achieved, a complex and widespread community of interest has been created. Significantly, it includes many key political and economic actors who express genuine commitment to addressing climate change. Effective and significant strategies to address global climate change will emerge when collective responsibility for finding solutions becomes a more widely accepted idea that underpins political decision-making and policy actions.

Second, we also derive optimism from the debates about what constitutes appropriate and timely responses to global climate change. Environmentalists and climate change activists taunt present leaders by suggesting that future generations will not forgive their inaction. It is possible that future generations will be less churlish because they may well hold more nuanced appreciations of the magnitude of the political challenges entailed in addressing climate change. They might appreciate the enormity of the work undertaken by the current generation of leaders who endeavour to create the informed publics needed for future action. At present, the most prudent path forward lies in judiciously embracing the precautionary principle. In itself, this is no small achievement because choosing to act with incomplete information requires accepting one kind of risk to avoid another (Kriebel et al., 2001, p. 875). Climate change mitigation policies may harm economic development, but over time these policy settings can be monitored, reviewed and revised to minimise these effects.

While we would like to believe that effective climate change responses might be achieved without significant costs and disruptions to current lifestyles and aspirations, this seems highly unlikely. The political dimensions of climate change responses instead suggest that disruptions are inevitable. Solutions to various social and economic problems have been achieved by political decisions and actions, and we believe that established mechanisms of international discussion and negotiation provide a framework through which new political solutions might be achieved. We therefore believe that global climate change can only be addressed by effective leadership and collective political will that enables new forms of political and economic organisation and new sources of political order.

Bibliography

Abi-Saab, G. (1997) *Human Rights and Humanitarian Law: The Quest for Universality*, Martinus Nijhoff, Boston, MA.
Achterberg, W. (2001) 'Environmental Justice and Global Democracy', in Gleeson, B. and Low, N. (eds.), *Governing for the Environment: Global Problems, Ethics and Democracy*, Palgrave, Houndmills.
Adger, N., Aggarwal, P., Agrawala, S., Joseph, A., Abdelkader, A., Cruz, R. V., De Alba Alcaraz, E., Easterling, W., Field, C., Fischlin, A., Fitzharris, B. B., García, C. G., Hanson, C., Harasawa, H., Huq, S., Jones, R., Bogataj, L. K., Karoly, D., Klein, R., Mortsch, L., Niang-Diop, I., Nicholls, R., Nováky, B., Nurse, L., Nyong, A., Oppenheimer, M., Palutikof, J., Parry, M., Patwardhan, A., Lankao, R., Rosenzweig, C., Schneider, S., Semenov, S., Smith, J. and Stone, J. (2007) *Climate Change 2007: Impacts, Adaptation and Vulnerability, Working Group II Contribution to the Intergovernmental Panel on Climate Change Fourth Assessment, Report Summary for Policymakers*, Cambridge University Press, Cambridge.
Adger, W. N. (2000) 'Social and Ecological Resilience: Are They Related?', *Progress in Human Geography*, Volume 24, Issue 3, pp. 347–364.
Adger, W. N. (2006) 'Vulnerability', *Global Environmental Change*, Volume 16, Issue 3, pp. 268–281.
Adger, W. N., Dessai, S., Goulden, M., Hulme, M., Lorenzoni, I., Nelson, D. R., Ness, L. O., Wolf, J. and Wreford, A. (2009) 'Are There Social Limits to Adaptation to Climate Change?', *Climatic Change*, Volume 93, Issue 3, pp. 335–354.
Adger, W. N., Huq, S., Brown, K., Conway, D. and Hulme, M. (2003) 'Adaptation to Climate Change in the Developing World', *Progress in Development Studies*, Volume 3, Issue 3, pp. 179–195.
Adger, W. N., Lorenzoni, I. and O'Brien, K. L. (eds.) (2009) *Adapting to Climate Change*, Cambridge University Press, Cambridge.
Africa Earth Observatory Network. Available at www.aeon.org.za/index/php (Accessed 13 March 2006).
Ahuja, D. and Tatsutani, M. (2009) 'Sustainable Energy for Developing Countries', *S.A.P.I.EN.S*, 2 January 2009, Online since 27 November 2009. Available at http://sapiens.revues.org/823 (Accessed 23 August 2012).
Alfsen, K. H., Sekeland, G. S. and Linnerud, K. (2010) 'Technological Change and the Role of Non-state Actors', in Biermann, F., Pattberg, P. and Zelli, F. (eds.), *Global Climate Governance Beyond 2012: Architecture, Agency and Adaptation*, Cambridge University Press, Cambridge.
Alley, R. B., Clark, P., Huybrechts, P. and Joughin, I. (2005) 'Ice-Sheet and Sea-Level Changes', *Science*, Volume 310, pp. 456–460.

Amstutz, M. R. (2005) *International Ethics: Concepts, Theories and Cases in Global Politics*, 2nd edition, Rowman and Littlefield, Oxford.

Anadon, L. D. and Holdren, J. P. (2009) 'Policy for Energy Technology Innovation', in Gallagher, K. S. (ed.), *Acting in Time on Energy Policy*, Brookings Institution Press, Washington, DC.

Anderson, B. (1991) *Imagined Communities*, Verso, London.

Anderson, T. L. and Leal, D. T. (1998) 'Visions of the Environment and Rethinking the Way We Think', in Dryzek, J. and Schlossberg, D. (eds.), *Debating the Earth*, Oxford University Press, Oxford.

Andersson, G., Donalek, P., Farmer, R., Hatziargyriou, N., Kamwa, I., Kundur, P., Martines, N., Paserba, J., Pourbeik, P., Sanchez-Gasca, J., Schulz, R., Stankovic, A., Taylor, C. and Vittal, V. (2005) 'Causes of the 2003 Major Grid Blackouts in North America and Europe, and Recommended Means to Improve System Dynamic Performance', *EIEE Transactions on Power Systems*, Volume 20, Issue 4, pp. 1922–1928.

Appiah, K. A. (2006) *Cosmopolitanism: Ethics in a World of Strangers*, Allen Lane, London.

Archibald, S., Roy, D. P., Van Wilgen, B. W. and Scholes, R. J. (2009) 'What Limits Fire? An Examination of Drivers of Burnt Area in Southern Africa', *Global Change Biology*, Volume 15, pp. 613–630.

Archibugi, D. (2001) 'The Politics of Cosmopolitan Democracy', in Gleeson, B. and Low, N. (eds.), *Governing for the Environment: Global Problems, Ethics and Democracy*, Palgrave, Houndmills.

Archibugi, D. and Held, D. (1995) *Cosmopolitan Democracy: An Agenda for a New World Order*, Polity, Cambridge.

Armaroli, N. and Balzani, V. (2007) 'The Future of Energy Supply: Challenges and Opportunities', *Angewandte Chemie International Edition*, Volume 46, Issue 1, pp. 52–66.

Arnold, D. G. (2011) 'Introduction: Climate Change and Ethics', in Arnold, D. G. (ed.), *The Ethics of Global Climate Change*, Cambridge University Press, Cambridge.

Arup, T. (2009) 'Australia, Indonesia in Carbon Trading Plan', *The Age*, 10 August. Available at http://www.theage.com.au/environment/australia-indonesia-in-carbon-trading-plan-20090809-ee9s.html (Accessed 14 November 2011).

Aubrey, C. (ed.) (2007) *Energy Revolution: A Sustainable Global Energy Outlook*, Greenpeace/European Renewable Energy Council (EREC). Available at http://www.energyblueprint.info/fileadmin/media/documents/energy_revolution.pdf (Accessed 21 August 2009).

Australian Broadcasting Corporation, *ABC Radio News 774*, Melbourne, VIC, 19 November 2009.

Barker, T., Bashmakov, I., Bernstein, L., Bogner, J. E., Bosch, P. R., Dave, R., Davidson, O. R., Fisher, B. S., Gupta, S., Halsnæs, K., Heij, G. J., Kahn Ribeiro, S., Kobayashi, S., Levine, M. D., Martino, D. L., Masera, O., Metz, B., Meyer, L. A., Nabuurs, G.-J., Najam, A., Nakicenovic, N., Rogner, H.-H., Roy, J., Sathaye, J., Schock, R., Shukla, P., Sims, R. E. H., Smith, P., Tirpak, D. A., Urge-Vorsatz, D. and Zhou, D. (2007) 'Technical Summary', in Metz, B., Davidson, O. R., Bosch, D. R., Dave, R. and Meyer, L. A. (eds.), *Climate Change 2007: Mitigation. Contribution of Working Group III to the Fourth Assessment Report of the Intergovernmental Panel on Climate Change*, Cambridge University Press, Cambridge and New York.

Barnett, J. (2001) *Security and Climate Change*, Tyndall Centre for Climate Change Research, Working Paper 7. Available at www.tyndall.ac.uk (Accessed 18 June 2012).

Barnett, J. (2009) 'The Prize of Peace (Is Eternal Vigilance): A Cautionary Editorial Essay on Climate Geopolitics', *Climatic Change*, Volume 96, Issues 1 and 2, pp. 1–6.

Barnett, M. F. (2004) *Rules for the World: International Organizations in World Politics*, Cornell University Press, Ithaca, NY.

Baron, J., Poff, N. L., Angermeier, P. L., Dahm, C. N., Gleik, P. H., Hairston, N. G., Jackson, R. B., Johnston, C. A., Richter, B. G. and Steinman, A. D. (2002) 'Meeting Ecological and Societal Needs for Freshwater', *Ecological Applications*, Volume 12, Issue 5, pp. 1247–1260.

Barry, B. (1986) 'Can States be Moral? International Morality and the Compliance Problem', in Ellis, A. (ed.), *Ethics and International Relations*, Manchester University Press, Manchester and Wolfeboro, NH.

Barry, J. and Eckersley, R. (2005) 'Whither the Green State?', in Barry, J. and Eckersley, R. (eds.), *The State and the Global Ecological Crisis*, The MIT Press, Cambridge, MA.

Bates, B., Kundezewicz, Z. W., Wu, S. and Palutikof, J. (eds.) (2008) *Climate Change and Water*, Technical Paper of the IPCC Secretariat, Intergovernmental Panel on Climate Change, Geneva.

Battisti, D. and Naylor, R. (2009) 'Historical Warnings of Future Food Insecurity with Unprecedented Seasonal Heat', *Science*, Volume 323, Issue 5922, pp. 240–244.

Baylis, J. (2008) 'The Concept of Security in International Relations', *Globalization and Environmental Challenges Hexagon Series on Human and Environmental Security and Peace*, Volume 3, Part V, pp. 495–502.

Bazerman, M. H. (2006) 'Climate Change as a Predictable Surprise', *Climatic Change*, Volume 77, Issue 1, pp. 179–193.

Bazerman, M. H. (2009) 'Barriers and Strategies for Overcoming Them', in Gallagher, K. S. (ed.), *Acting in Time on Energy Policy*, Brookings Institution Press, Washington, DC.

Bazerman, M. H. and Watkins, M. (2004) *Predictable Surprises: The Disasters You Should Have Seen Coming, and How to Prevent Them*, Harvard Business School Press, Boston, MA.

Bazilian, M., Hobbs, B. F., Blyth, W., MacGill, I. and Howells, M. (2011) 'Interactions Between Energy Security and Climate Change: A Focus on Developing Countries', *Energy Policy*, Volume 39, pp. 3750–3756.

BBC World Service Poll (2007) *Herald Sun*, 26 September 2007, p. 33.

Bearce, D. H., Floros, K. M. and McKibben, H. E. (2009) 'The Shadow of the Future and International Bargaining: The Occurrence of Bargaining in a Three-Phase Cooperative Framework', *The Journal of Politics*, Volume 17, Issue 2, pp. 719–732.

Bell, D. R. (2004) 'Environmental Refugees: What Right? Which Duties?', *Res Publica*, Volume 10, Issue 2, pp. 135–152.

Bell, R. G. (2006) 'What to Do About Climate Change', *Foreign Affairs*, Volume 85, Issue 3, pp. 105–113.

Biermann, F. (2005) 'The Rationale for a World Environment Organization', in Biermann, F. and Bauer, S. (eds.), *A World Environment Organization: Solution or Threat for Effective International Environmental Governance?*, Ashgate, Aldershot.

Biermann, F. and Bauer, S. (2005) 'The Debate on a World Environment Organization: An Introduction', in Biermann, F. and Bauer, S. (eds.), *A World Environment Organization: Solution or Threat for Effective International Environmental Governance?*, Ashgate, Aldershot.

Biermann, F. and Dingwerth, K. (2004) 'Global Environmental Change and the Nation State', *Global Environmental Politics*, Volume 4, Issue 1, pp. 1–22.

Biersteker, T. J. and Weber, C. (1996) 'The Social Construction of State Sovereignty', in Biersteker, T. J. and Weber, C. (eds.), *State Sovereignty as Social Construct*, Cambridge University Press, Cambridge.

Blatter, J. and Ingram, H. (2000) 'States, Markets and Beyond: Governance of Transboundary Water Resources', *Natural Resources Journal*, Volume 40, Issue 2, pp. 439–473.

Blatter, J. and Ingram, H. (2001) *Reflections on Water: New Approaches to Transboundary Conflicts and Cooperation*, The MIT Press, Cambridge, MA and London, UK.

Bodansky, D. (2001) 'The History of the Global Climate Change Regime', in Luterbacher, U. and Sprinz, D. F. (eds.), *International Relations and Global Climate Change*, The MIT Press, Cambridge, MA and London, UK.

Boehmer-Christiansen, S. (1996) 'The International Research Enterprise and Global Environmental Change: Climate-Change Policy as a Research Process', in Vogler, J. and Imber, M. (eds.), *The Environment and International Relations*, Routledge, London.

Bonanate, L. (1995) *Ethics and International Politics*, Polity, Cambridge.

Borgerson, S. (2008) 'Arctic Meltdown: The Economic and Security Implications of Global Warming', *Foreign Affairs*, Volume 87, Issue 2, pp. 63–77.

Brauch, H. G., Oswald Spring, U., Mesjasz, C., Grin, J., Dunay, P., Behara, N. C., Chourou, B., Kameri-Mbote, P. and Liotta, P. H. (eds.) (2008) *Global Environmental Challenges: Reconceptualising Security in the 21st Century*, Hexagon Series on Human and Environmental Security and Peace, Volume 3, Springer, Berlin, Heidelberg and New York.

Brennan, G. and Buchanan, J. M. (1985) *The Reason of Rules*, Cambridge University Press, New York.

British Broadcasting Company, *BBC News* 22 October 2008. Available at www.bbc.com/news (Accessed 30 October 2008).

Brown, L. (2005) 'A Planet Under Stress', in Dryzek, J. and Schlossberg, D. (eds.), *Debating the Earth*, 2nd edition, Oxford University Press, Oxford and New York.

Brownlie, I. (1998) *The Rule of Law in International Affairs: International Law at the Fiftieth Anniversary of the United Nations*, Martinus Nijhoff, The Hague.

Bull, H. (1995) *The Anarchical Society: A Study of Order in World Politics*, 2nd edition, Columbia University Press, New York.

Bull, H. (2002) *The Anarchical Society*, 3rd edition, Palgrave, Basingstoke and New York.

Burgess, J., Bedford, T., Hobson, K., Davies, G. and Harrison, C. (2003) '(Un)sustainable Consumption', in Berkhout, F., Leach, M. and Scoones, I. (eds.), *Negotiating Environmental Change: New Perspectives from Social Science*, Edward Elgar, Cheltenham.

Burroughs, W. (2003) *Climate: Into the 21st Century*, World Meteorological Organization, Cambridge University Press, Cambridge.

Buzan, B. (1991) *People, States and Fear: An Agenda for International Security Studies in the Post Cold-War Era*, 2nd edition, Harvester Wheatsheaf, New York.

Buzan, B. (1993) 'From International System to International Society: Structural Realism and Regime Theory Meet the English School', *International Organization*, Volume 47, Issue 3, pp. 327–352.

Cafaro, P. (2011) 'Beyond Business as Usual: Alternative Wedges to Avoid Catastrophic Climate Change and Create Sustainable Societies', in Arnold, D. G. (ed.), *The Ethics of Global Climate Change*, Cambridge University Press, Cambridge.

Cameron, R. (2008) 'Globalism as Structure: Ideas, Identities and Structural Change', *The Global Studies Journal*, Volume 1, Issue 4, pp. 29–38.

Camilleri, J. A., Malhotra, K. and Tehranian, M. (2000) *Reimagining the Future: Towards Democratic Governance*, Latrobe University, Melbourne, VIC.

Carcaillet, C., Bergman, I., Delorme, V., Hornberg, G. and Zackrisson, O. (2007) 'Long-term Fire Frequency Not Linked to Prehistoric Occupations in Northern Swedish Boreal Forest', *Ecology*, Volume 88, Issue 2, pp. 465–477.

Climate Science Watch (2008) 'Hurricanes and Climate Change: From the IPCC and Recent US Climate Change Science Program Reports', *Posted on Tuesday*, 2 September. Available at http://www.climatesciencewatch.org/index.php/csw/details/hurricanes_climate_ipcc_ccsp/ (Accessed 19 October 2008).

Cohen, T. (2008) 'How Britain Is the SIXTH Largest Water Importer in the World', *The Daily Mail*, 20 August. Available at www.dailymail.co.uk/news/article-1047158/Britain-sixth-largest-water-importer-world.html (Accessed 1 August 2010).

Collins, M. (2005) 'El NiÑo- or La NiÑa-like Climate Change?', *Climate Dynamics*, Volume 24, Issue 1, pp. 89–104.

Conca, K. (2005) 'Old States in New Bottles? The Hybridization of Authority in Global Environmental Governance', in Barry, J. and Eckersley, R. (eds.), *The State and the Global Ecological Crisis*, The MIT Press, Cambridge, MA.

Conway, J. (2009) *Cooperative Strategy for 21st Century Seapower*, Diane Publishing, Darby, PA.

Corson, W. H. (1994) 'Changing Course: An Outline of Strategies for a Sustainable Future', *Futures: the Journal of Forecasting and Planning*, Volume 26, Issue 2, pp. 206–223.

Cox, R. and Jacobson, H. K. (1997) 'A Framework for Inquiry', in Diehl, P. F. (ed.), *The Politics of Global Governance: International Organizations in an Interdependent World*, Rienner, Boulder, CO and London.

Dauncey, G. and Mazza, P. (2001) *Stormy Weather: 101 Solutions to Global Climate Change*, New Society Publishers, Gabriola Island.

Davies, A. (2007) 'Global Unity Needed for Climate Change: UN', *The Age*, 26 September. Available at http://www.theage.com.au/news/world/global-unity-needed-for-climate-change-un/2007/09/25/1190486308872.html (Accessed 18 June 2013).

Davis, S. T. and Caldeira, K. (2008) 'Consumption-based Accounting of CO_2 Emissions', *Proceedings of the National Academy of Sciences of the United States*, Volume 107, Issue 12, pp. 5687–5692.

de Almeida, P. and Silva, P. D. (2009) 'The Peak of Oil Production – Timings and Market Recognition', *Energy Policy*, Volume 37, Issue 4, Spring, pp. 1267–1276.

de Coninck, H., Fischer, C., Newell, R. G. and Ueno, T. (2007) 'International Technology-oriented Agreements to Address Climate Change', *Energy Policy*, Volume 36, Issue 1, pp. 335–356.

De Villiers, M. (2001) *Water Wars: Is the World's Water Running Out?*, Weidenfeld and Nicolson, London.

de Wit, M. and Stankiewicz, J. (2006) 'Changes in Surface Water Supply with Predicted Climate Change', *Science*, Volume 311, Issue 5769, pp. 1917–1921.

Deb, P. C., Singh, S. and Arora, D. R. (1983) *Green Revolution: An Assessment*, Beekay's Publications, Ludhiana.

Diehl, P. (ed.) (1997) *The Politics of Global Governance: International Organizations in an Interdependent World*, Lynne Rienner Publishers, London.

Diffenbaugh, N. and Scherer, M. (2011) 'Observational and Model Evidence of Global Emergence of Permanent, Unprecedented Heat in the 20th and 21st Centuries', *Climatic Change*, Volume 107, Issues 3 and 4, pp. 615–624.

Dimas, S. (2007) 'Big Carbon Cuts: Scary, but Doable', *csmonitor.com*. Available at http://www.csmonitor.com/2007/0927/p09s01-coop.html (Accessed 25 March 2009).

DiMento, J. F. C. and Doughman, P. (2007a) 'Climate Change: How the World is Responding', in DiMento, J. F. C. and Doughman, P. (eds.), *Climate Change: What It Means for Us, Our Children and Our Grandchildren*, The MIT Press, Cambridge, MA and London, UK.

DiMento, J. F. C. and Doughman, P. (2007b) 'Climate Change: What It Means to Us, Our Children and Grandchildren', in DiMento, J. F. C. and Doughman, P. (eds.), *Climate Change: What It Means for Us, Our Children and Our Grandchildren*, The MIT Press, Cambridge, MA and London, UK.

Donnelly, J. (2000) *Realism and International Relations*, Cambridge University Press, Cambridge.

Doppelt, G. (1978) 'Walzer's Theory of Morality in International Relations', *Philosophy and Public Affairs*, Volume 8, Issue 1, pp. 3–26.

Doty, R. L. (1996) 'Sovereignty and the Nation: Constructing the Boundaries of National Identity', in Biersteker, T. J. and Weber, C. (eds.), *State Sovereignty as Social Construct*, Cambridge University Press, Cambridge.

Dow, K. and Downing, T. E. (2006) *The Atlas of Climate Change: Mapping the World's Greatest Challenge*, University of California Press, Berkeley, CA.

Dow, K. and Downing, T. E. (2007) *The Atlas of Climate Change: Mapping the World's Greatest Challenge*, University of California Press, Berkeley, CA.

Dryzek, J. (1997) *Politics of the Earth: Environmental Discourses*, Oxford University Press, New York.

Dryzek, J. and Schlossberg, D. (eds.) (1998) *Debating the Earth*, Oxford University Press, Oxford and New York.

Dryzek, J. and Schlossberg, D. (eds.) (2005) *Debating the Earth*, 2nd edition, Oxford University Press, Oxford and New York.

Dube, O. P. (2009) 'Linking Fire and Climate: Interactions with Land Use, Vegetation and Soil', *Current Opinion in Environmental Sustainability*, Volume 1, Issue 2, pp. 161–169.

Dunne, T. (2001a) 'Sociological Investigations: Instrumental, Legitimist and Coercive Interpretations of International Society', *Millennium: Journal of International Studies*, Volume 30, Issue 1, pp. 67–91.

Dunne, T. (2001b) 'New Thinking on International Society', *British Journal of Politics and International Relations*, Volume 3, Issue 2, June, pp. 223–244.

Dyer, H. C. (1996) 'Environmental Security as a Universal Value: Implications for International Theory', in Vogler, J. and Imber, M. F. (eds.), *The Environment and International Relations*, Global Environmental Change Programme, Routledge, London and New York.

Eakin, H., Tompkins, E. L., Nelson, D. R. and Anderies, J. M. (2009) 'Hidden Costs and Disparate Uncertainties: Trade-offs in Approaches to Climate Policy', in Adger, W. N., Lorenzoni, I. and O'Brien, K. L. (eds.), *Adapting to Climate Change: Thresholds, Values and Governance*, Cambridge University Press, Cambridge.

Eckersley, R. (1992) *Environmentalism and Political Theory: Towards an Ecocentric Approach*, UCL Press, London.

Eckersley, R. (2004) *The Green State: Rethinking Democracy and Sovereignty*, The MIT Press, Cambridge, MA.

Eckersley, R. (2005) 'Greening the Nation-State: From Exclusive to Inclusive Sovereignty', in Barry, J. and Eckersley, R. (eds.), *The State and the Global Ecological Crisis*, The MIT Press, Cambridge, MA.

Eckersley, R. (2006) 'Communitarianism', in Dobson, A. and Eckersley, R. (eds.), *Political Theory and the Ecological Challenge*, Cambridge University Press, Cambridge.

Edmondson, E. (2001) 'The Intergovernmental Panel on Climate Change: Beyond Monitoring?', in Gleeson, B. and Low, N. (eds.), *Governing for the Environment: Global Problems, Ethics and Democracy*, Palgrave, Houndmills.

Edmondson, B. (2008) 'Global Order: Accommodating Diversity in the 21st Century', *The Global Studies Journal*, Volume 1, Issue 2, pp. 25–34.

Edmondson, B. (2009) 'The Impossible Dream: Consensus-Based International Climate Change Responses', *The Global Studies Journal*, Volume 2, Issue 3, pp. 1–14.

Edmondson, B. (2011) 'Collective Responsibilities: New Principles for Order in the 21st Century', *The Global Studies Journal*, Volume 3, Issue 4, pp. 11–20.

Edmondson, B. and Levy, S. (2008) *International Relations: Nurturing Reality*, Pearson Education, Frenchs Forest.

Ellwood, D. T. (2009) 'Foreword', in Gallagher, K. S. (ed.), *Acting in Time on Energy Policy*, The Brookings Institution Press, Washington, DC.

El-Raey, M. (1997) 'Vulnerability Assessment of the Coastal Zone of the Nile Delta of Egypt, to the Impacts of Sea Level Rise', *Ocean and Coastal Management*, Volume 37, Issue 1, pp. 29–40.

Evans, A. (2010) 'Resource Scarcity, Climate Change and the Risk of Violent Conflict', *World Development Report 2011, Background Paper*. Available at http://siteresources.worldbank.org/EXTWDR2011/Resources/6406082-1283882418764/WDR_Background_Paper_Evans.pdf (Accessed 7 May 2012).

Evenson, R. and Gollin, D. (2003) 'Assessing the Impact of the Green Revolution, 1960 to 2000', *Science*, Volume 300, Issue 7, pp. 758–762.

Evett, R., Franco-Vizcaino, E. and Stephens, S. (2007) 'Comparing Modern and Past Fire Regimes to Assess Changes in Prehistoric Lightning and Anthropogenic Ignitions in a Jeffrey Pine – Mixed Conifer Forest in the Sierra San Pedro Mártir, Mexico', *Canadian Journal of Forest Research*, Volume 37, Issue 2, pp. 318–330.

Farrands, C. (1996) 'Society, Modernity and Social Change: Approaches to Nationalism and Identity', in Krause, J. and Renwick, N. (eds.), *Identities in International Relations*, Macmillan with St Antony's College, Oxford.

Finnemore, M. (1996) *National Interests in International Society*, Cornell University Press, New York.

Flannery, T. (2005) *The Weather Makers: The History and Future Impact of Climate Change*, Text Publishing, Melbourne.

Fowler, M. R. and Bunck, J. M. (1995) *Law, Power and the Sovereign State, The Evolution and Application of the Concept of Sovereignty*, The Pennsylvania State University Press, Pennsylvania.

Friedman, E. J., Hochstetler, K. and Clark, A. M. (2005) *Sovereignty, Democracy and Global Civil Society: State–Society Relations at UN World Conferences*, State University of New York Press, New York.

Friedman, G. and Starr, H. (1997) *Agency, Structure, and International Politics: From Ontology to Empirical Inquiry*, Routledge, London and New York.

Friel, S., Marmot, M., McMichael, A. J., Kjellstrom, T. and Vågerö, D. (2008) 'Global Health Equity and Climate Stabilisation: A Common Agenda', *The Lancet*, Volume 372, Issue 9650, pp. 1677–1683.

Frumkin, H., Hess, J., Luber, G., Malilay, J. and McGeehin, M. (2008) 'Climate Change: The Public Health Response', *American Journal of Public Health*, Volume 98, Issue 3, pp. 435–445.

Fukuyama, F. (1992) *The End of History and the Last Man*, Hamish Hamilton, London.

Gallagher, K. S. (2009) 'Acting in Time on Climate Change', in Gallagher, K. S. (ed.), *Acting in Time on Energy Policy*, The Brookings Institution Press, Washington, DC.

Gardiner, S. M. (2011) *A Perfect Moral Storm: The Ethical Challenge of Climate Change*, Oxford University Press, Oxford.

Gardiner, S. M. and Hartzell-Nichols, L. (2012) 'Ethics and Global Climate Change', *Nature Education Knowledge*, Volume 3, Issue 10, pp. 5–11.

Garnaut, R. (2011) *The Garnaut Review 2011: Australia in the Global Response to Climate Change*, Cambridge University Press, Melbourne, VIC.

Gelbspan, R. (2001) 'Foreword', in Dauncey, G. and Mazza, P. (eds.), *Stormy Weather: 101 Solutions to Global Climate Change*, New Society Publishers, Gabriola Island.

Gilboa, E. (2005) 'The CNN Effect: The Search for a Common Theory of International Relations', *Political Commons*, Volume 22, Issue 1, pp. 27–44.

Giordano, M. A. and Wolf, A. T. (2003) 'Sharing Waters: Post-Rio International Water Management', *Natural Resources Forum*, Volume 27, Issue 2, pp. 163–171.

Girardet, H. (2000) 'Greening Urban Society', in Fox, W. (ed.), *Ethics and the Built Environment: Professional Ethics*, Routledge, New York.

Gleeson, B. and Low, N. (eds.) (2001) *Governing for the Environment: Global Problems, Ethics and Democracy*, Palgrave, Houndmills.

Gleick, P. H. (1993) 'Water and Conflict: Fresh Water Resources and International Security', *International Security*, Volume 18, Issue 1, pp. 79–112.

Gleik, P. H. et al. (2010) 'Climate Change and the Integrity of Science', Letters (Sills, J. ed.), *Science*, Volume 328, Issue 5979, pp. 689–690. Available at http://www.sciencemag.org/content/328/5979/689.full (Accessed 8 September 2011).

Gore, A. (2006) *An Inconvenient Truth: The Planetary Emergency of Global Warming and What We Can Do About It*, Rodale Press, Emmaus.

Granström, A. and Niklasson, M. (2008) 'Potentials and Limitations for Human Control Over Historic Fire Regimes in the Boreal Forest', *Philosophical Transactions of the Royal Society B Biological Sciences*, Volume 363, Issue 1501, pp. 2353–2358.

Grissino-Mayer, H. D., Romme, W. H., Floyd, M. L. and Hanna, D. D. (2004) 'Climatic and Human Influences on Fire Regimes of the Southern San Juan Mountains, Colorado, USA', *Ecology*, Volume 85, Issue 6, pp. 1708–1724.

Gupta, J. (2005) 'Global Environmental Governance: Challenges for the South from a Theoretical Perspective', in Biermann, F. and Bauer, S. (eds.), *A World Environment Organization: Solution or Threat for Effective International Environmental Governance?*, Ashgate, Aldershot.

Haas, P. M. (1990) 'Obtaining International Environmental Protection Through Epistemic Consensus', *Millennium: Journal of International Studies*, Volume 19, Issue 3, pp. 347–363.

Haas, P. M. (1995) 'Epistemic Communities and the Dynamics of International Environmental Cooperation', in Rittberger, V. with the assistance of Mayer, P. (ed.), *Regime Theory and International Relations*, Clarendon Press, Oxford.

Haas, P. M. (1997) 'Introduction: Epistemic Communities and International Policy Coordination', in Haas, P. M. (ed.), *Knowledge, Power, and International Policy Coordination*, University of South Carolina Press, Columbia, SC.

Haas, P. M., Keohane, R. O. and Levy, M. A. (eds.) (1993) *Institutions for the Earth: Sources of Effective International Environmental Protections*, The MIT Press, Cambridge, MA.

Hajat, S. and Kosatky, T. (2010) 'Heat-related Mortality: A Review and Exploration of Heterogeneity', *Journal of Epidemiology & Community Health*, Volume 64, Issue 9, pp. 753–760.

Halliday, F. (1994) *Rethinking International Relations*, Macmillan Press, New York.

Hammitt, J. K. (2000) 'Global Climate Change: Benefit–Cost Analysis vs. the Precautionary Principle', *Human and Ecological Risk Assessment*, Volume 6, Issue 3, pp. 387–398.

Hasenclever, A., Mayer, P. and Rittberger, V. (1997) *Theories of International Regimes*, Cambridge University Press, New York.

Hashmi, S. H. (1997) *State Sovereignty, Change and Persistence in International Relations*, Pennsylvania University Press, Pennsylvania.

Haughton, G. (2007) 'In Pursuit of the Sustainable City', in Marcotullio, P. J. and McGranahan, G. (eds.), *Scaling Urban Environmental Challenges: From Local to Global and Back*, Earthscan, London.

Hawken, P. (2005) 'The Next Industrial Revolution', in Dryzek, J. and Schlossberg, D. (eds.), *Debating the Earth*, 2nd edition, Oxford University Press, Oxford and New York.

Held, D. (2010) 'Reframing Global Governance: Apocalypse Soon or Reform!', in Brown, G. W. and Held, D. (eds.), *The Cosmopolitanism Reader*, Polity, Cambridge and Malden, MA.

Held, D., McGrew, A., Goldblatt, D. and Perraton, J. (1999) *Global Transformations: Politics, Economics and Culture*, Polity Press, Cambridge.

Hempel, L. C. (1996) *Environmental Governance: the Global Challenge*, Island Press, Washington, DC.
Henkin, L. (1989) 'The Role of Law and its Limitations', in Williams, M. (ed.), *International Relations in the Twentieth Century: A Reader*, MacMillan Education, London.
Herz, J. H. (1957) 'Rise and Demise of the Territorial State', *World Politics*, Volume 9, Issue 4, pp. 473–493.
Heyd, T. and Brooks, N. (2009) 'Exploring Cultural Dimensions to Climate Change', in Adger, W. N., Lorenzoni, I. and O'Brien, K. L. (eds.), *Adapting to Climate Change: Thresholds, Values, Governance*, Cambridge University Press, Cambridge.
Higgott, R., Underhill, G. R. D. and Bieler, A. (2000) *Non-state Actors and Authority in the Global System*, Routledge, London.
Hoffman, J. and Hoffman, M. (2008) *Green: Your Place in the New Energy Revolution*, Palgrave Macmillan, Houndmills.
Hoffman, S. (1997) 'Foreword', in Hashmi, S. H. (ed.), *State Sovereignty, Change and Persistence in International Relations*, Pennsylvania State University Press, Pennsylvania.
Hoffman, S. (2002) 'Foreword to the Second Edition: Revising the Anarchical Society', in Bull, H. (ed.), *The Anarchical Society*, 3rd edition, Palgrave, New York, pp. xxiv–xxix.
Hoffmann, W. A., Schroeder, W. and Jackson, R. B. (2002) 'Positive Feedbacks of Fire, Climate, and Vegetation and the Conversion of Tropical Savannah', *Geophysical Research Letters*, Volume 29, Issue 22, pp. 2052–2056.
Hogan, W. (2009) 'Electricity Market Structure and Infrastructure', in Gallagher, K. S. (ed.), *Acting in Time on Energy Policy*, Brookings Institution Press, Washington, DC.
Holden, B. (2000) *Global Democracy: Key Debates*, Routledge, London.
Holdren, J. P. (2006) 'The Energy Innovation Imperative: Addressing Oil Dependence, Climate Change and Other 21st Century Energy Challenges', *Innovations, Technology, Governance, Globalization*, Volume 1, Issue 2, pp. 3–23.
Holgate, S. (2007) 'On the Decadal Rates of Sea Level Change During the Twentieth Century', *Geophysical Research Letters*, Volume 34, L01602, doi:10.1029/2006GL028492.
Hollins, L. (2010) *Dealing with Change, Complexity and Uncertainty in Global Climate Change Negotiations – is Adaptiveness the Answer?*, M.Sc. Thesis, Lund University. Available at http://www.lumes.u.se/database/alumni/08.10/Thesis/Hollins_Livia_Thesis2.pdf (Accessed 14 February 2012).
Holsti, O. R. (1995) 'Models of International Relations: Realist and Neoliberal Perspectives on Conflict and Cooperation', in Kegley, C. W. Jr. and Wittkopf, E. R. (eds.), *The Global Agenda, Issues and Perspectives*, 4th edition, McGraw-Hill, New York.
Homer-Dixon, T. (1994) 'Environmental Scarcities and Violent Conflict: Evidence from Cases', *International Security*, Volume 19, Issue 1, Summer, pp. 5–40.
Houghton, J. J. (2004) *Global Warming: The Complete Briefing*, 3rd edition, Cambridge University Press, Cambridge.
Huggett, R. J. (2006) *The Natural History of the Earth: Debating Long-term Change in the Geosphere and Biosphere*, Routledge, New York and London.
Hughes, J. D. (2009) *The Environmental History of the World: Humankind's Changing Role in the Community of Life*, Routledge, New York.

Hunold, C. and Dryzek, J. (2005) 'Green Political Strategy and the State: Combining Political Theory and Comparative History', in Eckersley, R. and Barry, J. (eds.), *The State and the Global Ecological Crisis*, The MIT Press, Cambridge, MA.
Huntington, S. P. (1993) 'The Clash of Civilizations?', *Foreign Affairs*, Volume 72, Issue 3, Summer, pp. 22–49.
Huntington, T. (2006) 'Evidence for Intensification of the Global Water Cycle: Review and Synthesis', *Journal of Hydrology*, Volume 319, Issue 1–4, pp. 83–95.
Hurrell, A. (1995) 'International Society and the Study of Regimes: A Reflective Approach', in Rittberger, V. (ed.), *Regime Theory and International Relations*, Oxford University Press, Oxford.
Hurrell, A. (2007) *On Global Order: Power, Values and the Constitution of International Society*, Oxford University Press, Oxford and New York.
Inayatullah, N. and Blaney, D. L. (2004) *International Relations and the Problem of Difference*, Routledge, New York and London.
Intergovernmental Panel on Climate Change (2007a) 'Summary for Policymakers', in Solomon, S., Qin, D., Manning, M., Chen, Z., Marquis, M., Averyt, K. B., Tignor, M. and Miller, H. L. (eds.), *Climate Change 2007: The Physical Science Basis. Contribution of Working Group 1 to the Fourth Assessment Report of the Intergovernmental Panel on Climate Change*, Cambridge University Press, Cambridge and New York.
Intergovernmental Panel on Climate Change (2007b) *Climate Change 2007: Synthesis Report. Contribution of Working Groups I, II and III to the Fourth Assessment Report of the Intergovernmental Panel on Climate Change*, IPCC, Geneva, Switzerland.
International Food Policy Research Institute (2002) *Green Revolution, Curse of Blessing*. Available at http://www.ifpri.org/sites/default/files/pubs/pubs/ib/ib11.pdf (Accessed 2 September 2011).
International Food Policy Research Institute (2009) *Climate Change – Impacts on Agriculture and the Costs of Adaptation*. Available at http://www.sciencealert.com.au/features/20090810-19957.html (Accessed 6 September 2011).
International Union for Conservation of Nature (2011) 'Plenty More Fish in the Sea? Not for Much Longer', 19 April. Available at http://iucn.org/?7268/Plenty-more-fish-in-the-sea-Not-for-much-longer (Accessed 30 October 2012).
Jackson, R., Carpenter, S., Dahm, C., McKnight, D., Naiman, R., Postel, S. and Running, S. (2001) 'Water in a Changing World', *Ecological Applications*, Volume 11, Issue 4, pp. 1027–1045.
Jackson, R. and Sorensen, G. (2007) *Introduction to International Relations: Theories and Approaches*, 3rd edition, Oxford University Press, Oxford.
Jacobsson, S. and Johnson, A. (2000) 'The Diffusion of Renewable Energy Technologies: An Analytical Framework and Key Issues for Research', *Energy Policy*, Volume 28, Issue 9, pp. 625–640.
Johansson, D. J. and Azar, C. (2003) 'Analysis of Land Competition Between Food and Bioenergy', *World Resources Review*, Volume 15, Issue 2, pp. 165–175.
Jolly, R. and Ray, D. B. (2007) 'Human Security – National Perspectives and Global Agendas: Insights from National Human Development Reports', *Journal of International Development*, Volume 19, Issue 4, pp. 457–472.
Kandji, S., Verchot, L. and Mackensen, J. (2006) 'Climate Change and Variability in the Sahel Region: Impacts and Adaptation Strategies in the Agricultural

Sector', *United Nations Environmental Programme and World Agroforestry Center*, Nairobi.

Kanellos, M. (2008) 'Scientific Group: Cut Carbon Dioxide Emissions in Half', *CNET News*, Green Tech, 24 January. Available at http://news.cnet.com/8301-11128_3-9857531-54.html (Accessed 25 March 2009).

Kanter, J. and Revkin, A. (2007) 'World Scientists Near Consensus on Warming', *New York Times*, 29 January. Available at http://www.nytimes.com/2007/01/30/world/30climate.html (Accessed 6 September 2011).

Karmad, B. (2008) 'Security and Sovereignty', *Globalization and Environmental Challenges Hexagon Series on Human and Environmental Security and Peace*, Volume 3, Part IV, pp. 421–430.

Kasperson, R. E. and Kasperson, J. X. (2001) *Climate Change, Vulnerability and Social Justice*, Risk and Vulnerability Programme, Stockholm Environment Institute, Stockholm.

Kegley, C. W. Jr. and Raymond, G. A. (2005) *The Global Future: A Brief Introduction to World Politics*, Thomson/Wadsworth, Belmont, CA.

Keohane, R. O. and Nye, J. S. (2001) *Power and Interdependence*, 3rd edition, Longman, New York.

Kjéllen, B. (2006) 'Foreword', in Dow, K. and Downing, T. E. (eds.), *The Atlas of Climate Change: Mapping the World's Greatest Challenge*, University of California Press, Berkeley, CA.

Koremenos, B. (2005) 'Contracting around International Uncertainty', *American Political Science Review*, Volume 99, Issue 4, pp. 549–565.

Krakoff, S. (2011) 'Parenting the Planet', in Arnold, D. G. (ed.), *The Ethics of Global Climate Change*, Cambridge University Press, Cambridge.

Krasner, S. (1982a) 'Structural Causes and Regime Consequences: Regimes as Intervening Variables', *International Organization*, Volume 36, Issue 2, pp. 185–205.

Krasner, S. (1982b) 'Regimes and the Limits of Realism: Regimes as Autonomous Variables', *International Organization*, Volume 36, Issue 2, pp. 497–510.

Krasner, S. (1999) *Sovereignty: Organized Hypocrisy*, Princeton University Press, Princeton, NJ.

Kratochwil, F. V. (1989) *Rules, Norms and Decisions: On the Conditions of Practical and Legal Reasoning in International Relations and Domestic Affairs*, Cambridge University Press, Cambridge.

Krause, J. and Renwick, N. (eds.) (1996) *Identities in International Relations*, St. Antony's Series, Macmillan/St. Martin's Press, Houndmills and New York.

Kriebel, D., Tickner, J., Epstein, P., Lemons, J., Levins, R., Loechler, E. L., Quinn, M., Rudel, R., Schettler, T. and Stoto, M. (2001) 'The Precautionary Principle in Environmental Science', *Environmental Health Perspectives*, Volume 109, Issue 9, pp. 871–876.

Kukreja, S. (2005) 'The Two Faces of Development', in Balaam, D. N. and Veseth, M. (eds.), *Introduction to International Political Economy*, Pearson Prentice Hall, Upper Saddle River, NJ.

Kütting, G. (2000) *Environment, Society and International Relations: Towards More Effective International Environmental Agreements*, Routledge, London.

Laslett, P. (2001) 'Environmental Ethics and the Obsolescence of Existing Political Institutions', in Gleeson, B. and Low, N. (eds.), *Governing for the Environment: Global Problems, Ethics and Democracy*, Palgrave, Houndmills.

Leaver, R. and Richardson, J. L. (eds.) (1993) *The Post-Cold War Order: Diagnoses and Prognoses*, Allen and Unwin, St. Leonards, NSW.

Lee, H. (2009) 'Oil Security and the Transportation Sector', in Gallagher, K. S. (ed.), *Acting in Time on Energy Policy*, Brookings Institution Press, Washington, DC.

Lieserowicz, A. (2006) 'Climate Change Risk Perception and Policy Preferences: The Role of Affect, Imagery and Values', *Climatic Change*, Volume 77, Issues 1 and 2, pp. 46–72.

Lind, R. C. (1995) 'Intergenerational Equity, Discounting and the Role of Cost-benefit Analysis in Evaluating Climate Policy', *Energy Policy*, Volume 23, Issue 4, pp. 379–389.

Linden, E. (2006) *The Winds of Change: Climate, Weather, and the Destruction of Civilizations*, Simon & Schuster, New York.

Lipson, C. (1991) 'Why Are Some International Agreements Informal?', *International Organization*, Volume 45, Issue 4, pp. 495–538.

Low, N. and Gleeson, B. (1998) *Justice, Society and Nature: An Exploration of Political Ecology*, Routledge, London.

Low, N. and Gleeson, B. (2001) 'The Challenge of Ethical Environmental Governance', in Gleeson, B. and Low, N. (eds.), *Governing for the Environment: Global Problems, Ethics and Democracy*, Palgrave, Houndmills.

Luard, E. (1990) *International Society*, The MacMillan Press, Hampshire.

Luterbacher, U. and Sprinz, D. F. (2001) 'Problems of Global Environmental Cooperation', in Luterbacher, U. and Sprinz, D. F. (eds.), *International Relations and Global Climate Change*, The MIT Press, Cambridge, MA and London, UK.

Manheim, J. B. (1984) 'Changing National Images: International Public Relations and Media Setting Agenda', *The American Political Science Review*, Volume 7, Issue 3, pp. 641–657.

Manne, A. and Richels, R. (1995) 'The Greenhouse Debate: Economic Efficiency, Burden Sharing and Hedging Strategies', *The Energy Journal*, Volume 16, Issue 4, pp. 1–38.

Manning, C. W. (1962) *The Nature of International Society*, G. Bell and Sons, London.

Mansbach, R. and Rafferty, K. L. (2008) *Introduction to Global Politics*, Routledge, London and New York.

Markowitz, E. and Shariff, A. (2012a) 'The Moral Case of Climate Change', *Climate Science and Policy*, pp. 1–3. 26 September. Available at www.climatescienceandpolicy.eu/2012/09/the-moral-case-of-climate-change/ (Accessed 18 June 2013).

Markowitz, E. and Shariff, A. (2012b) 'Climate Change and Moral Judgement', *Nature Climate Change*, Volume 2, pp. 243–247, doi:10.1038/nclimate1378.

Marland, G. and Obersteiner, M. (2008) 'Large-scale Biomass for Energy, with Considerations and Cautions: An Editorial Comment', *Climatic Change*, Volume 87, Issues 3 and 4, pp. 335–348.

Mason, C. (2003) *The 2030 Spike: Countdown to Global Catastrophe*, Earthscan Publications, London, UK and Sterling, VA.

Matthew, R. (2007) 'Climate Change and Human Security', in DiMento, J. F. C. and Doughman, P. (eds.), *Climate Change: What It Means for Us, Our Children, and Our Grandchildren*, The MIT Press, Cambridge, MA.

Mathews, J. T. (ed.) (1991) *Preserving the Global Environment: The Challenge of Shared Leadership*, W. W. Norton, New York.

McCartney, G. and Hanlon, P. (2008) 'Climate Change and Rising Energy Costs: A Threat but also an Opportunity for a Healthier Future?', *Public Health: Journal of the Royal Institute of Public Health*, Volume 122, Issue 7, pp. 653–656.

Mekonnen, M. M. and Hoekstra, A. Y. (2011) 'National Water Footprint Accounts: The Green, Blue and Grey Water Footprint of Production and Consumption', *Value of Water Research Report Series No.50*, UNESCO-IHE, Delft, the Netherlands.

Mendez, R. P. (1999) 'Peace as a Global Public Good', in Kaul, I., Grunberg, I. and Stern, M. A. (eds.), *Global Public Goods: International Cooperation in the 21st Century*, Oxford University Press, United Nations Development Programme (UNDP), Oxford.

Metz, B., Davidson, O. R., Bosch, P. R., Dave, R. Meyer. L. A., (2007) *Climate Change 2007: Mitigation. Contribution of Working Group III to the Fourth Assessment Report of the Intergovernmental Panel on Climate Change*, Cambridge University Press, Cambridge and New York.

Minion, J. M., Kinsella, W. J., O'Neill, C. and Peterson, T. R. (2009) 'New Media, New Movements?', in Endres, D., Sprain, L. M. and Peterson, T. R. (eds.), *Social Movements to Address Climate Change: Local Steps for Global Action*, Cambria Press, New York.

Mirza, M. M. Q., Warrick, R. A. and Ericksen, N. J. (2003) 'The Implications of Climate Change on Floods of the Ganges, Brahmaputra and Meghna Rivers in Bangladesh', *Climate Change*, Volume 57, Issue 3, pp. 287–318.

Mische, P. (1989) 'Ecological Security and the Need to Reconceptualize Sovereignty', *Alternatives: Global, Local, Political*, Volume 14, Issue 4, pp. 389–427.

Morgenthau, H. J. (1973) *Politics Among Nations: The Struggle for Power and Peace*, 5th edition, Alfred A Knopf Inc, New York.

Morgenthau, H. J. (2005) *Politics Among Nations: The Struggle for Power and Peace*, 7th edition, McGraw Hill, Boston, MA.

Morse, E. L. and Myers, A. (2001) 'Strategic Energy Policy Challenges for the 21st Century', *Report of an Independent Task Force, Sponsored by the James A. Baker III Institute for Public Policy of Rice University and the Council on Foreign Relations*, Council on Foreign Relations. Available at www.cfr.org/energy-security/strategic-energy-policy-challenges-21st-centuiry/p3942 (Accessed 4 May 2009).

Mosher, D. (2011) 'Gates: "Cute" Tech Won't Solve Planet's Energy Woes', *Wired Business Conference: Disruptive by Design*, 3 May 2011. Available at http://www.wired.com/epicenter/2011/05/bill-gates-energy-tech/ (Accessed 15 February 2012).

Mulligan, S. (2010) 'Energy, Environment, and Security: Critical Links in a Post-Peak World', *Global Environmental Politics*, Volume 10, Issue 4, pp. 79–100.

Murphy, A. (1996) 'The Sovereign State System as Political-territorial Ideal: Historical and Contemporary Considerations', in Biersteker, T. J. and Weber, C. (eds.), *State Sovereignty as Social Construct*, Cambridge University Press, Cambridge.

Nadelmann, E. A. (1990) 'Global Prohibition Regimes: The Evolution of Norms in International Society', *International Organization*, Volume 44, Issue 4, Autumn, pp. 479–526.

Najam, A. (2005a) 'Neither Necessary, Nor Sufficient: Why Organizational Tinkering Will Not Improve Environmental Governance', in Biermann, F. and Bauer, S. (eds.), *A World Environment Organization: Solution or Threat for Effective International Environmental Governance?*, Ashgate, Aldershot.

Najam, A. (2005b) 'Developing Countries and Global Environmental Governance: From Contestation to Participation to Engagement', *International Environmental Agreements*, Volume 5, pp. 303–321.

Nardin, T. (1983) *Law, Morality and the Relations of States*, Princeton University Press, Princeton, NJ.

Nardin, T. (2008) 'Theorising the International Rule of Law', *Review of International Studies*, Volume 34, Issue 3, pp. 385–401.

Nelson, D. R. (2009) 'Conclusions: Transforming the World', in Adger, W. N., Lorenzoni, I. and O'Brien, K. L. (eds.), *Adapting to Climate Change: Thresholds, Values, Governance*, Cambridge University Press, Cambridge.

Newman, P., Beatley, T. and Boyer, H. (2009) *Resilient Cities: Responding to Peak Oil and Climate Change*, Island Press, Washington, DC.

Nicholls, R. J. (2003) *Case Study on Sea-level Rise Impacts*, OECD Workshop on the Benefits of Climate Policy: Improving Information for Policy Makers, Working Party on Global and Structural Policies.

Nicholls, R. J. and Mimura, N. (1998) 'Regional Issues Raised by Sea Level Rise and their Policy Implications', *Climate Research*, Volume 11, December, pp. 5–18.

Nicholls, R. J. and Tol, R. (2006) 'Impacts and Responses to Sea-Level Rise: A Global Analysis of the SRES Scenarios over the Twenty-First Century', *Philosophical Transactions of the Royal Society A: Mathematical, Physical and Engineering Sciences*, Volume 364, Issue 1841, pp. 1073–1095.

Nicholls, R. J., Hanson, S., Herweijer, C., Patmore, N., Hallegatte, S., Corfee-Morlot, J., Chateau, J. and Muir-Wood, R. (2007a) *Ranking Port Cities with High Exposure and Vulnerability to Climate Extremes – Exposure Estimates, OECD Environmental Working Paper No. 1*, Organisation for Economic Co-operation and Development (OECD), Paris.

Nicholls, R. J., Wong, P. P., Burkett, V. R., Codignotto, J. O., Hay, J. E., McLean, R. F., Ragoonaden, S. and Woodroffe, C. D. (2007b) 'Coastal Systems and Low-lying Areas', in Parry, M. L., Canziani, O. F., Palutikof, J. P., van der Linden, P. J. and Hanson, C. E. (eds.), *Climate Change 2007: Impacts, Adaptation and Vulnerability. Contribution of Working Group II to the Fourth Assessment Report of the Intergovernmental Panel on Climate Change*, Cambridge University Press, Cambridge, pp. 315–356.

Nyberg, D. (1993) *The Varnished Truth*, University of Chicago Press, Chicago, IL.

O'Brien, K. L. and Leichenko, R. M. (2000) 'Double Exposure: Assessing the Impacts of Climate Change within the Context of Economic Globalisation', *Global Environmental Change*, Volume 10, Issue 3, pp. 221–232.

O'Neill, K. (2009) *The Environment and International Relations*, Cambridge University Press, Port Melbourne, VIC.

Oberthür, S. O. and Ott, H. (1999) *The Kyoto Protocol: International Climate Policy for the 21st Century*, Springer, Berlin.

Ohmae, K. (1991) *The Borderless World, Power and the Strategy in the Interlinked Economy*, Fontana, London.

Ohmae, K. (1995) *The End of the Nation State, The Rise of Regional Economies*, HarperCollins Publishers, London.

Okin, G. S., Parsons, A. J., Wainwright, J., Herrick, J. E., Bestelmeyer, B. T., Peters, D. C. and Fredrickson, E. L. (2009) 'Do Changes in Connectivity Explain Desertification?', *Bioscience*, Volume 59, Issue 3, pp. 237–244.

Overpeck, J. T. and Cole, J. E. (2007) 'Climate Change: Lessons from a Distant Monsoon', *Nature*, Volume 445, January, pp. 270–271.

Owen, R. (2008) 'World Food Supply Must Rise 50%, Ban Ki Moon Tells Rome Summit', *Times Online*, 3 June. Available at http://www.timesonline.co.uk/tol/news/world/europe/article4056801.ece (Accessed 2 September 2011).

Oxfam (2007) 'Adapting to Climate Change – What's Needed in Poor Countries, and Who Should Pay', *Oxfam Briefing Paper 104*, Oxfam International, Oxford, UK. Available at http://www.oxfam.org/sites/www.oxfam.org/files/adapting to climate change (Accessed 6 September 2011).

Oye, K. A. (2005) 'The Conditions for Cooperation in World Politics', in Oye, K. A. (ed.), *International Politics: Enduring Concepts and Contemporary Issues*, Pearson Longman, New York.

Page, E. (2006) *Climate Change, Justice and Future Generations*, Edward Elgar, Cheltenham, UK.

Parry, M. L., Canziani, O. F., Palutikof, J. P., van der Linden, P. J. and Hanson, C. E. (eds.) (2007) *Climate Change 2007: Impacts Adaption and Vulnerability. Contribution of Working Group II to the Fourth Assessment Report of the Intergovernmental Panel on Climate Change*, Cambridge University Press, Cambridge.

Pasternak, A. D. (2000) *Global Energy Futures and Human Development: A Framework for Analysis*, U.S. Department of Energy, Lawrence Livermore National Laboratory, National Technical Information Service, Springfield, VA.

Paterson, M. (2000) *Understanding Global Environmental Politics: Domination, Accumulation and Resistance*, St. Martin's Press, New York.

Patz, A., Engelberg, D. and Last, J. (2000) 'The Effects of Changing Weather on Public Health', *Annual Review of Public Health*, Volume 21, pp. 271–307.

Pausas, J. G. (2004) 'Changes in Fire and Climate in the Eastern Iberian Peninsula (Mediterranean Basin)', *Climatic Change*, Volume 63, Issue 3, pp. 337–350.

Pearce, F. (2007) *With Speed and Violence: Why Scientists Fear Tipping Points in Climate Change*, Beacon Press, Boston, MA.

Pearce, F. (2009) 'How 16 Ships Create as Much Pollution as All the Cars in the World'. Available at http://www.dailymail.co.uk/sciencetech/article-1229857/How-16-ships-create-pollution-cars-world.html (Accessed 2 September 2011).

Pennington, M. (2001) 'Environmental Markets vs. Environmental Deliberation: A Hayekian Critique of Green Political Economy', *New Political Economy*, Volume 6, Issue 2, pp. 171–190.

Peters, S. (2003) 'The Shortcomings of Western Response Strategies to New Energy Vulnerabilities', *Energy Exploration & Exploitation*, Volume 21, Issue 1, pp. 29–60.

Philpott, D. (1997) 'Ideas and the Evolution of Sovereignty', in Hashmi, S. H. (ed.), *State Sovereignty, Change and Persistence in International Relations*, The Pennsylvania State University Press, University Park, PA.

Philpott, D. (2001) *Revolutions in Sovereignty*, Princeton University Press, Princeton, NJ.

Pincen, T. and Finger, M. (1994) *Environmental NGOs in World Politics*, Routledge, London.

Pinderhughes, R. (2004) *Alternative Urban Futures: Planning for Sustainable Development in Cities Throughout the World*, Bowman and Littlefield, Lanham, MD.

Pogge, T. (1992) 'Cosmopolitanism and Sovereignty', *Ethics*, Volume 103, Issue 1, pp. 48–75.

Ponting, C. (1991) *A Green History of the World*, Sinclair-Stevenson, London.

Postel, S. (2000) 'Entering an Era of Water Scarcity: The Challenges Ahead', *Ecological Applications*, Volume 10, Issue 4, pp. 941–948.
Postiglione, A. (2001) 'An International Court of the Environment', in Gleeson, B. and Low, N. (eds.), *Governing for the Environment: Global Problems, Ethics and Democracy*, Palgrave, Houndmills.
Power, M. J., Marlon, J., Ortiz, N., Bartlein, P. J., Harrison, S. P., Mayle, F. E., Ballouche, A., Bradshaw, R. H. W., Carcaillet, C. and Cordova, C. (2008) 'Changes in Fire Regimes since the Last Glacial Maximum: An Assessment Based on a Global Synthesis and Analysis of Charcoal Data', *Climate Dynamics*, Volume 30, Issues 7 and 8, pp. 887–907.
Prugh, T., Flavin, C. and Sawin, J. L. (2008) 'Changing the Oil Economy', in Starke, L. (ed.), *State of the World 2008: Global Security*, Worldwatch Institute/Earthscan, London.
Rabkin, J. A. (2005) *Law Without Nations: Why Constitutional Government Requires Sovereign States*, Princeton University Press, Princeton, NJ.
Raustiala, K. (2001) 'Nonstate Actors in the Global Climate Regime', in Luterbacher, U. and Sprinz, D. F. (eds.), *International Relations and Global Climate Change*, The MIT Press, Cambridge, MA and London, UK.
Renaud, F., Julca, A., Warner, K., Maza, M. and Oliver-Smith, A. (2010) 'Climate Change, Environmental Degradation and Migration', *Natural Hazards*, Volume 55, Issue 3, pp. 689–715.
Reus-Smit, R. (1997) 'The Constitutional Structure of International Society and the Nature of Fundamental Institutions', *International Organisation*, Volume 51, Issue 4, Autumn, pp. 555–589.
Richardson, J. L. (1993) 'The End of Geopolitics?', in Leaver, R. and Richardson, J. L. (eds.), *The Post-Cold War Order: Diagnoses and Prognoses*, Allen and Unwin, St. Leonards, NSW.
Rittberger, V. (1995) 'Editor's Introduction', in Rittberger, V. with the assistance of Mayer, P. (ed.), *Regime Theory and International Relations*, Clarendon Press, Oxford.
Rittberger, V. with the assistance of Mayer, P. (ed.) (1993) *Regime Theory and International Relations*, Oxford University Press, New York.
Roaf, S., Crichton, D. and Nicol, F. (2005) *Adapting Buildings and Cities for Climate Change: A 21st Century Survival Guide*, Architectural Press, Amsterdam and Boston, MA.
Roberts, J. T. and Parks, B. C. (2007) *A Climate of Justice: Global Inequality, North-south Politics and Climate Policy*, The MIT Press, Cambridge, MA.
Rogers, P. P. and Leal, S. (2010) *Running Out of Water: The Looming Crisis and Solutions to Conserve Our Most Precious Resource*, Palgrave Macmillan, New York.
Romero-Lankao, P. R. (2008) *Urban Areas and Climate Change: Review of Current Issues and Trends*, Issues Paper for the 2011 Global Report on Human Settlements, National Centre for Atmospheric Research. Available at www.isse.ucar.edu/romero-lankao/GRHS_2011_IssuesPaperfinal.pdf (Accessed 5 January 2012).
Rowlands, I. H. (2001) 'Classical Theories of International Relations', in Luterbacher, U. and Sprinz, D. F. (eds.), *International Relations and Global Climate Change*, The MIT Press, Cambridge, MA and London, UK.
Rudolph, C. (2005) 'Sovereignty and Territorial Borders in a Global Age', *International Studies Review*, Volume 7, Issue 1, pp. 1–20.

Sachs, J. D. (2005) 'The Geography of Economic Development', in Oatley, T. (ed.), *The Global Economy: Contemporary Debates*, Pearson Longman, New York.

Sawin, J. (2003) 'Charting a New Energy Future', in Worldwatch Institute, *State of the World 2003: Progress Towards a Sustainable Society*, Worldwatch Institute/Earthscan, London.

Sax, J. L. (1990) 'The Constitution, Property Rights and the Future of Water Law', *University of Colorado Law Review*, Volume 61, Issue 2, pp. 257–282.

Scambos, T., Hulbe, C. and Fahnestock, M. (2003) 'Climate-induced Ice Shelf Disintegration in the Antarctic Peninsula', in Domack, E., Levente, A., Burnet, A., Bindschadler, R., Convey, P., and Kirby, M. (eds.), *Paleobiology and Paleoenvironments of Eoscene Rocks, Antarctic Research Series*, Volume 79, pp. 79–92, American Geophysical Union, Washington DC. Available at http://www.agu.org/books/ar/v079/AR079p0079/AR079p0079.shtml (Accessed 17 September 2010).

Schachter, O. (1991) 'The Emergence of International Environmental Law', *Journal of International Affairs*, Volume 44, Issue 2, Winter, pp. 457–493.

Scharpf, F. W. (1989) 'Decision Rules, Decision Styles and Policy Choices', *Journal of Theoretical Politics*, Volume 1, Issue 2, pp. 149–176.

Schmidt, M. G. (1989) *Common Heritage or Common Burden*, Clarendon, Oxford.

Schrag, D. P. (2009) 'Making Carbon Capture and Storage Work', in Gallagher, K. S. (ed.), *Acting in Time on Energy Policy*, Brookings Institution Press, Washington, DC.

Schumacher, E. F. (1974) *Small Is Beautiful: A Study of Economics as if People Mattered*, Abacus, London.

Science and Environmental Health Network (2000) *The Precautionary Principle: A Common Sense Way to Protect Public Health and the Environment*. Available at http://www.mindfully.org/Precaution/Precautionary-Principle-Common-Sense.htm (Accessed 16 September 2011).

Sebenius, J. K. (1992) 'Challenging Conventional Explanations of International Cooperation: Negotiation Analysis and the Case of Epistemic Communities', *International Organization*, Volume 46, Issue 1, Winter, pp. 323–365.

Shapin, S. (1994) *A Social History of Truth*, University of Chicago Press, Chicago, IL.

Serageldin, I. (2010) 'Water Wars? A Talk with Ismail Serageldin', *World Policy Journal*, Volume 26, Issue 4, pp. 25–31.

Sherwood, S. C. and Huber, M. (2010) 'An Adaptability Limit to Climate Change Due to Heat Stress', *Proceedings of the National Academy of Sciences*, Volume 107, Issue 21, pp. 9552–9555.

Shrader-Frechette, K. S. (1998) *Environmental Ethics*, Rowman & Littlefield Publishers Inc., Lanham, MD.

Siegart, F., Ruecker, G., Hinrichs, A. and Hoffmann, A. (2001) 'Increased Damage from Fires in Logged During Draughts Caused by El Niño', *Nature*, Volume 414, November, pp. 437–440.

Smith, K. and Petley, D. N. (2009) *Environmental Hazards: Assessing Risk and Reducing Disaster*, Routledge, Milton Park and New York.

Snow, D. M. (2006), *Cases in International Relations: Portraits of the Future*, 2nd edition, Pearson/Longman, New York.

Sprain, L., Peterson, N., Vickery, M. and Schutten, J. K. (2009) 'Environment 2.0: New Forms of Social Activism', in Endres, D., Sprain, L. M. and Peterson, T. R. (eds.), *Social Movements to Address Climate Change: Local Steps for Global Action*, Cambria Press, New York.

Sprinz, D. F. and Weiβ, M. (2001) 'Domestic Politics and Global Climate Policy', in Luterbacher, U. and Sprinz, D. F. (eds.), *International Relations and Global Climate Change*, The MIT Press, Cambridge, MA and London, UK.

Stalley, P. (2003) 'Environmental Scarcity and International Conflict', *Conflict Management and Peace Science*, Volume 20, Issue 2, pp. 33–58.

Stern, H, (2006) *Stern Review on the Economics of Climate Change*. Available at http://webarchive.nationalarchives.gov.uk/+/http:/www.hm-treasury.gov.uk/sternreview_index.htm (Accessed 23 July 2009).

Stern, N. (2009) *A Blueprint for a Safer Planet: How to Manage Climate Change and Create a New Era of Progress and Prosperity*, The Bodley Head, London.

Stirling, A. (2007) 'Risk, Precaution and Science: Towards a More Constructive Policy Debate', *European Molecular Biology Organization (EMBO) Reports*, Volume 8, Issue 4, pp. 309–315.

Stokke, O. S. and Vidas, D. (1996) *Governing the Antarctic: The Effectiveness and Legitimacy of the Antarctic Treaty System*, Cambridge University Press, Cambridge and New York.

Strzepek, K. and Boehlert, B. (2010) 'Competition for Water for the Food System', *Philosophical Transactions of the Royal Society Biological Sciences*, Volume 365, Issue 1554, pp. 2927–2940.

Suhrke, A. (1994) 'Environmental Degradation and Population Flows', *Journal of International Affairs*, Volume 47, Issue 2, pp. 473–496.

Syphard, A. D., Radeloff, V. C., Keeley, J. E., Hawbaker, T. J., Clayton, M. K., Stewart, S. I. and Hammer, R. B. (2007) 'Human Influence on California Fire Regimes', *Ecological Applications*, Volume 17, Issue 5, pp. 1388–1402.

Tannenwald, N. and Wohlforth, W. C. (2005) 'Introduction: The Role of Ideas and the End of the Cold War', *Journal of Cold War Studies*, Volume 7, Issue 2, pp. 3–12.

The Commission on Global Governance (1995) *Our Global Neighbourhood*, Oxford University Press, Oxford.

Thompson, J. (2001) 'Planetary Citizenship: The Definition and Defence', in Gleeson, B. and Low, N. (eds.), *Governing for the Environment: Global Problems, Ethics and Democracy*, Palgrave, Houndmills.

Tilman, D., Cassman, K. G., Maston, P. A., Naylor, R. and Polasky, S. (2002) 'Agricultural Sustainability and Intensive Production Practices', *Nature*, Issue 418, pp. 671–677.

Tilman, D. and Lehman, C. (2001) 'Human-caused Environmental Change: Impacts on Plant Diversity and Evolution', *Proceedings of the National Academy of Sciences of the United States of America*, Volume 98, Issue 10, pp. 5433–5440.

Toffler, A. (1971) *Future Shock*, The Bodley Head, London.

Transboundary Freshwater Dispute Database. Available at http://www.transboundarywaters.orst.edu/index.html (Accessed 8 September 2011).

Tubiello, F. N., Soussana, J.-F. and Howden, S. M. (2007) 'Crop and Pasture Response to Climate Change', *Proceedings of the National Academy of Sciences of the United States of America*, Volume 104, Issue 50, pp. 19686–19690.

Turton, H. and Barreto, L. (2006) 'Long-term Security of Energy Supply and Climate Change', *Energy Policy*, Volume 34, Issue 15, pp. 2232–2250.

Umbach, F. (2010) 'Global Energy Security and the Implications for the EU', *Energy Policy*, Volume 38, Issue 3, pp. 1229–1240.

United Nations (1945a) *United Nations Charter and Statute of the International Court of Justice*. Available at http://www.un.org/en/documents/charter (Accessed 4 June 2013).

United Nations (1945b) *United Nations Charter*. Available at www.un.org/en/documents/charter/ (Accessed 6 December 2011).

United Nations (1999) *Executive Board of the United Nations Development Programme and the United Nations Population Fund First Regular Session*, 25–29 January, New York.

United Nations Committee for Development Policy (2009) *Achieving Sustainable Development in an Age of Climate Change*, Policy Note, United Nations Publishing Section, New York.

United Nations Development Programme (2008) *Human Development Report 2007/2008 Fighting Climate Change: Human Solidarity in a Divided World*, National Human Development Report Unit. Available at http://hdr.undp.org/en/reports (Accessed 23 July 2012).

United Nations Educational, Scientific and Cultural Organization (2005a) 'UNESCO Water e-Newsletter No. 92: Water Use'. Available at http://www.unesco.org/water/news/newsletter/92.shtml (Accessed 8 September 2011).

United Nations Educational, Scientific and Cultural Organization (2005b) 'World Commission on the Ethics of Scientific Knowledge and Technology (COMEST)', *The Precautionary Principle*, UNESCO, Paris.

United Nations Educational, Scientific and Cultural Organization (2009) 'United Nations World Water Development Report 3', *Water in a Changing World*, World Water Assessment Programme. Available at www.unesco.org/new/en/natural-sciences/environment/water/wwap/wwdr/wwdr3/2009 (Accessed 2 April 2012).

United Nations Environment Programme (2005) *One Planet, Many People: Atlas of Our Changing Environment*, EarthPrint.com.

United Nations Framework Convention on Climate Change (2011) *Kyoto Protocol, Status of Ratification*. Available at http.unfcc,int/kyoto_protocol/status_of_ratification/items/2613.php (Accessed 1 December 2011).

United Nations General Assembly (1970) *Declaration on Principles of International Law Concerning Friendly Relations and Cooperation Among States in Accordance with the Charter of the United Nations*, Resolution 25/2625, Geneva. Available at www.un-documents.net/a25r2625.htm (Accessed 14 September 2011).

United Nations General Assembly (1988) *Resolution for Protection of Global Climate for Present and Future Generations of Mankind Resolution 43/53*, 70th Plenary Meeting, 6 December, Geneva. Available at www.un.org/documents/ga/res/43/a43r053.htm (Accessed 4 October 2011).

United Nations Human Settlements Programme (2008) *State of the World's Cities 2008/9: Harmonious Cities*, UN-HABITAT, Earthscan, London.

Utting, P. (2000) *Business Responsibility for Sustainable Development*. United Nations Research Institute for Social Development, Occasional Paper No. 2, Geneva.

van der Werf, G. R., Randerson, J. T., Giglio, L., Gobron, N. and Dolman, A. J. (2008) 'Climate Controls on the Variability of Fires in the Tropics and Subtropics', *Global Biogeochemical Cycles*, Volume 22, GB3028, doi:10.1029/2007GB003122.

Van Langervelde, F., Van De Vijver, C. A. D. M., Kumar, L., Van de Koppel, J., De Ridder, N., Van Andel, J., Skidmore, A. K., Hearne, J. W., Strrojnijder, L. and Bond, W. J. (2003) 'Effects of Fire and Herbivory on the Stability of Savanna Ecosystems', *Ecology*, Volume 84, Issue 2, pp. 337–350.

Van Noorden, R. (2008) 'Global Industries Call for Carbon Cuts', 20 June. Available at http://www.rsc.org/chemstryworld/News/2008/June/20060801.asp (Accessed 25 March 2009).

van Oldenborgh, G. J., Philip, S. Y. and Collins, M. (2005) 'El Niño in a Changing Climate: A Multi-model Study', *Ocean Science: An Interactive Open Access Journal of the European Geosciences Union*, Volume 1, Issue 2, pp. 81–95.

Veblen, T. T., Kitzberger, T. and Donnegan, J. (2000) 'Climatic and Human Influences on Fire Regimes in Ponderosa Pine Forests in the Colorado Front Range', *Ecological Applications*, Volume 10, Issue 4, pp. 1178–1195.

Venton, C. C. (2007) *Climate Change and Water Resources*, WaterAid, Environmental Resources Management, London.

Vitousek, P., Mooney, H., Lubchenco, J. and Melillo, J. (1997) 'Human Domination of Earth's Ecosystems', *Science*, Volume 277, July, pp. 494–499.

Vogler, J. (1996) 'Introduction: The Environment in International Relations: Legacies and Contentions', in Vogler, J. and Imber, M. (eds.), *The Environment and International Relations*, Routledge, London.

Vogler, J. (2005) 'In Defense of International Environmental Cooperation', in Barry, J. and Eckersley, R. (eds.), *The State and the Global Ecological Crisis*, The MIT Press, Cambridge, MA.

Vogler, J. and Imber, M. (1996) *The Environment and International Relations*, Routledge, New York.

Vörösmarty, C., Green, P., Salisbury, J. and Lammers, R. (2000) 'Global Water Resources: Vulnerability from Climate Change and Population Growth', *Science*, Volume 289, Issue 284, pp. 284–288.

Vörösmarty, C. and Sahagian, D. (2000) 'Anthropogenic Disturbance of the Terrestrial Water Cycle', *Bioscience*, Volume 50, Issue 9, pp. 753–765.

Wackernagel, M., Schulz, N., Deumling, D., Linares, A., Jenkins, M., Kapos, V., Monfreda, C., Loh, J., Myers, N., Norgaard, R. and Randers, J. (2002) 'Tracking the Ecological Overshoot of the Human Economy', *Proceedings of the National Academy of Sciences of the United States of America*, Volume 99, Issue 14, pp. 9266–9271.

Walker, R. B. J. (1991) 'State Sovereignty and the Articulation of Political Space/Time', *Millennium: Journal of International Studies*, Volume 20, Issue 3, pp. 445–461.

Waltz, K. N. (1979) *Theory of International Politics*, Addison Wesley, Reading, MA.

Walzer, M. (2000) *Just and Unjust Wars: A Moral Argument with Historical Illustrations*, Basic Books, New York.

Wang, X. and McAllister, R. (2011) 'Adapting to Heatwaves and Coastal Flooding', in Cleugh, H., Smith, M., Battaglia, M. and Graham, P. (eds.), *Climate Change Science and Solutions for Australia*, CSIRO Publishing, Collingwood, VIC, Australia.

Weisman, A. (2004) 'Mining the Imagination for New Energy: Scientists Call for a Research Blitz Targeting Extreme Possibilities', *The Los Angeles Times*, 25 July. Available at http://articles.latimes.com/2004/jul25/opinion/op-weisman25 (Accessed 8 December 2008).

Weiss, T. (2011) *Thinking About Global Governance: Why People and Ideas Matter*, Routledge, Oxon.

Weltzin, J. F., Loik, M. E., Schwinning, S., Williams, D. G., Fay, P. A., Haddad, B. M., Harte, J., Husman, T. E., Knapp, A. K., Lin, G., Pockman, W. T., Shaw, M. R., Small, E. E., Smith, M. D., Smith, S. D., Tissue, D. T. and Zak, J. C. (2003) 'Assessing the Response of Terrestrial Ecosystems to Potential Changes in Precipitation', *Bioscience*, Volume 53, Issue 10, pp. 941–952.

Wendt, A. (2005) 'Anarchy Is What States Make of It', in Oye, K. A. (ed.), *International Politics: Enduring Concepts and Contemporary Issues*, Pearson Longman, New York.

Williams, M. (ed.) (1989) *International Relations in the Twentieth Century: A Reader*, MacMillan Education, London.

Wilson, P. (2008) 'Russia the Next Climate Recalcitrant', *The Australian*, Worldwide 10, 17 November.

Wolf, A. T. (1998) 'Conflict and Cooperation Along International Waterways', *Water Policy*, Volume 1, Issue 2, pp. 251–265.

World Wildlife Fund (2008) *Living Planet Report*. Available at http://www.wwf.org.uk/wwf_articles.cfm?unewsid=2294 (Accessed 22 July 2009).

World Wildlife Fund (2012) *Living Planet Report 2012*, World Wildlife Fund in Conjunction with Global Footprint Network and ZSL Living Conservation. Available at http://awsassets.panda.org/downloads/1_lpr_2012_online_full_size_single_pages_final_120516.pdf (Accessed 31 August 2012).

Yergin, D. (2006) 'Ensuring Energy Security', *Foreign Affairs*, Volume 85, Issue 2, pp. 69–82.

Yohe, G., Andronova, N. and Schlesinger, M. (2004) ' "To Hedge or Not Against Climate Change" Policy Forum', *Science*, Volume 306, Issue 5695, pp. 416–417.

Young, O. (1989) *International Cooperation: Building Regimes for Natural Resources and the Environment*, Cornell University Press, Ithaca, NY.

Young, O. (1994) *International Governance: Protecting the Environment in a Stateless Society*, Cornell University Press, Ithaca, NY.

Young, O. (1997) *Global Governance: Drawing Insights from the Environmental Experience*, The MIT Press, Cambridge, MA.

Young, O. (2002) *The Institutional Dimensions of Environmental Change: Fit, Interplay, and Scale*, The MIT Press, Cambridge, MA.

Zhao, M. and Running, S. (2010) 'Drought-induced Reduction in Global Terrestrial Net Primary Production from 2000 Through 2009', *Science*, Volume 329, Issue 5994, pp. 940–943.

Index

Note: Bold locators denotes figure and table

accountability, 34, 63, 185, 187, 191, 203
acid rainfall, 27, 206
activists, 160, 171, 175, 226
adaptation policies, 12–13, 130, 203
adaptation strategies, 17–19, 54, 56, 88, 93, 95, 101–2, 187–8, 190–3, 195–7, 203
advocacy, 71
affluent states, 15, 80, 81, 83
Africa, 30, 55, 82, 88, 93, 148
agricultural land, 45, 51, 143, 150, 177
agricultural production, 23, 52, 84, 90, 107, 124, 140, 145, 186, 209
agriculturally productive, 47, 90, 189
agriculture, 19, 84, 87–8, 107, 152–3, 220
allegiances, 177
anthropogenic emissions, 23, 88, 166
aquaculture, 135, 144, 151
Asia, 55, 148, 150
aspirations, 60, 66, 69, 87–8, 96–7, 99, 105–6, 114, 117–20, 179, 205, 216–17, 219, 221–2, 225–6
atmospheric temperatures, 133–5, 142, 144, 148, 152, 189, 198, 224
attitudes, 20, 95, 108, 207, 209, 220
Australia, 88, 104, 138
autonomous security, 19–20

balance of power, 3, 114, 202
bargaining, 8, 14, 31, 168, 170, 193, 203, 213
biodiversity, 220
biofuels, 124
biomass, 124, 137
borders, 8–11, 18, 23–4, 45, 60–1, 64, **75**, 92, 106, 121, 153, 162, 176–7, 179–80, 184, 192–3, 195–7

capitalism, 33, 66, 85, 109
capitalist economies, 53
carbon dioxide, 17, 23, 103, 118, 132
carbon tax/taxes, ix, 2, 81, 104, 223
carbon trading, 63, 135–6, 223
characteristics of states, 37, 187
China, 88, 143, 209
cities, 16, 35, 82–3, 91–2, 108, 116, 121, 137, 141, 143, 151–3, 173
citizens, 9–10, 24–5, 33–5, **56**–8, 73–4, 79–81, 173, 199–200, 210–11, 213–14
clean energy, 3, 52
climate change impacts, 13–14, 34–5, 41–2, 82, 84, 98, 138–9, 169–70, 173, 179–80, 189–90, 211–12, 222
climatic patterns, 17, 34, 66, 88, 147, 153
coastal, 9, 19, **23**, 82, 91, 93, 140–**4**
Cold War, 31, 160–1
collective action(s), 8, 12–13, 47, 64, 69, 81, 189, 225
collective responsibilities, 44, 56, 58, 189, 193–5, 203, 224–5
collective security, 5, 6, 16, 30, 37, 48, 56, 58, 67, 75, 77, 126, 129–32, 178, 184, **200**
common good, 12, 46, 200, 203
compensation, 27, 140
compliance, 34, 37–8, 41–2, 61–3, 65, 99, 167
conflict management, 89
conflicts, vii, 17, 33, 75, 84, 90, 101, 123, 132, 160, 183
consensus, 3–4, 14, 36–8, 40–1, 43–4, 49, 67, 129, 158, 172, 178, 201
consumption, 4, 50–1, 54–5, 81–3, 85–7, 96–100, 104–8, 110–14, 118–22, 126–8, 131–2, 135–6, 145, 217

249

conventions, 7–8, 38, 42, 52, 64, 70, 158, 199, 212, 214
Copenhagen Summit, 14–15, 38–9, 89, 99, 189, 190–1
corporations, 6, 24, 67, 78, 80, 108, 122, 184, 210
crisis, 28, 32, 34, 140, 167, 190, 211, 220
customary law, 37

data, 7, 22, 84
democracy, ix, 1, 22, 44, 51, 68, 78, 84, 219
democratic principles, 6, 70, 191
democratizing, 29, 178
desertification, 83, 93, 148, 153, 175
developed states, 22, 44, 88, 92, 96–9, 110–13, 116–18, 145, 178, 223
developing states, 9, 78, 82–4, 88, 93–4, 96–9, 105–7, 114, 116–17, 119–21, 124–5, 223
diplomacy, 16, 29, 48, 77, 130, 162, **200**
diseases, 26, **136**, 148–50, 190
dispute settlement, 93
disruption, 95, 107, 112, 132, 152, 176
diversification, 115, 119, 125
droughts, 16, 22, 26, 80, 90, 93, 98, 139–41, 141, 148, 150–2
dust, 90, **136**, 153

East–West, 218
ecological security, 185, 215
economic actors, 71–2, 78–9, 104–5, 107, 109, 111–2, 125–6, 176, 191–2, 194–5, 220, 223
economic challenges, 61, 89, 204, 210
economic costs, 67, 104
economic order, viii, 55, 142
economic policies, 33, 46, 192, 206
economic progress, 19, 21, 29, 51–2, 68, 103, 106, 197
economic prosperity, 14, 19–20, 34–5, 45, 53–5, **62–3**, 77, 96, 118, 120–1, 132, 187, 219, 222
economics, 60, 83, 85, 87, 126, 221–2
economic systems, 3, 52, 102, 182, 207, 222

ecosystem, 9, 11, 18, 21, 77, 90, 137, 195–6, 198, 209, 216
effective leadership, 98, 188, 203, 226
effective policies, 126, 196
electoral popularity, 34, 101, 223
electoral systems, 75
energy consumption, 4, 26, 50–1, 82, 86, 96, 99–100, 104, 107–8, 113, 119, 127, 190
energy costs, 115–6
energy policies, 33, 46, 192, 206
energy production, 4, 49, 98, 100, 102, 104–5, 108, 117–19, 121–4, 126, 130–2, 223–4
energy systems, 25, 115, 129
enforcement, 27, 41, 61–2, 65, 106, 177, 211
environmental management, 20, 205
environmental organisations, 171
equity, 27, 55, 64, 112, 126–7, 222
equitable, ix, 2, 4, 13, 50, 53, 64, 87, 97, 103, 153, 193, 208–9, 216
ethical decision-making, 167, 211, 213
ethical considerations, 127, 212
ethics, **200**, 205, 209, 212–3
Europe, 36, 88, 93, 138, 140, 145
expertise, 2, 5, 78–9, 161, 164–5, 167, 169, 179
extinctions, 140, 190, 206, 208, 215
extreme weather events, 8, 18, 34, 125, 138

fair/fairness, 212, 215
false dichotomy, 81, 212, 215–16
fires, 34, 90, 135, 141, 146–8, 154, 189
floods, 16, 34, 80, 90, 93, **136**–7, 139–41, 147–52
food production, 16, 18, 26, 82, 84, 107–8, 134, 136–7
food security, 8, **23**, 84, 107, 140, 145, 148
food trade, 140
fossil fuel, 18, 25, 66, 97–9, 101, 111, 113, 124, 126, 132, 156, 190, 209
framework, viii, 28, 69, 72, 105, 129, 166, 171, 214, 22, 226
freedom, 12, 74, 76, 141, 210–11
freshwater, 17, 89–91, 137, **143**, 148–9, 152

Index 251

functional capacities, 45–6, 198, 213
future generations, ix, 2, 69, 121, 175, 179, 205, 210, 214, 226

Ganges River, 36, 148
glacial, **23**, 148–9
GDP, 12, 114, 145
geographic borders, 8, 43, 64
geophysical systems, 63, 168, 189
global commons, 77, 185, 194, 208–9, 212, 214, 217
global population, 8, 19, 26–7, 82, 97, 107, 117, 119, 144, 205, 210
global warming, 1, 25, 129, 140, 166
global water cycle, 134, 136
governance mechanisms, 18, 50–1, 58, 115, 130, 165, 170, 203, 206, 209, 224
government ownership, 122
governmental capacities, 65, 222
green, 28, **72**, 84–5, 107, 124, 166, 217
greenhouse gas emissions, 15, 18, 24–5, 49, 63, 81, 98–9, 101, 104, 107, 116, 118, 120, 127, 129–30, 134, 152, 219, 223
greenhouse gases, vi, 12, 22, 41, 44, 99, 116, 134, 167, 173–4, 178
Greenpeace, 71, **72**, 168

habitable land, 26, 34, 143, **144**
habitability, 4, 52, 96, 180, 193, 224
happiness, 55, 141
health risks, 140, 149
hedging, 127
higher rainfalls, 81
human rights, 8, 58, 65, 72–3, 90, 142, 178, 185, 212, 214–15
human settlements, 19, 52, 135
hydroelectricity, 92, 135, 140, 150
hydrological cycle, 84, 88, 147, 151–2

ice, 27, 134, 136, 139, 142, 148
ideas of progress, 159, 206
ideology/ideologies, 17, 45, 129, 160, 196, 217
IGOs, 6, 69, **72**, **75**
irrigation, 52, 90, 92, 135, 140, 156
impasses, 85, 96
incremental gains, 109

independent authority, **75**, 163, 174, 179, 212
industrialisation, 22, 31, 88–91, 96–7, 99, 101, 120, 125, 156, 172–4, 205, 209, 211
industrialised production, 52, 65, 88, 94
industrial progress, 11, 104, 156, 187
infrastructure, 9, 54, 80–2, 88, 92, 108–10, 114, 116–18, 122, 129, 138, 143, 145, 147, 150, 151, 153, 221
innovation, 84, 86, 109, 113, 157
insurance, 34, 109, 140, **144**, 146
integration, 25, 168, 221
interdependence, 5, 30, 43, 46, 54, 59, 63, 180, 201, 203
interdependency, 5, 31
interest groups, 71, 73, **75**
Intergovernmental Panel on Climate Change, 9, 22, 41
Intergovernmental organisations, 6–7, 15, 39–40, 65, 69, 105, 202, 212
international agreements, 18–19, 34, 37–8, 41–2, 45, 47, 50, 53, 61, 69, 89–90, 102–3, 105, 130, 163, 174, 190, 197, 199–200, 203–4, 211–14, 223
international civil society, 24, **75**
international consensus, 41, 44
international institutions, viii, 7, 10, 13, 30, 69, 160, 163, 195, 204, 214–15
international law, 6, 27, 30, 38, 47, 58, **75**, 93, 164, 184, 211–12
international norms, 7, 37, 164, 224
international order, ix, 7–8, 10, 12, 17, 29, 77–8, 83, 120, 190, 200–3, 212
international organisations, 8, 48, **59**, 76, 102, 125–6, 165, 170, 195, 205, 212
international political economy, 15, 58, 85–6, 119, 142, 157, 187, 197, 220, 221
intervention, 6, 16, 19–20, 27, 56, 65, 77, 111, 123, 130

jurisdiction, 179, 183
justice, 18, 64, 70, 209

knowledge based actors, 40
knowledge communities, 165–8, 170–1
Kyoto Protocol, 40–1, 69, 89, 103, 129, 174, 184, 213

land use, 22, 146
Law of the Sea, 37, 53, 146
legal equality, 10, 46, 63, **75**, 162, 183, 201, 211
legislation, 42, 47, 50, 63, 119, 212
legitimacy, 7, 32, 38, 44, 58, **75**–6, 78–9, 181, 186, 194, 198, 225
legitimate authority, 64, 73–4, 155, 193
liberal democracies, 33, 35, 53, 56–8, 62, 101, 174
linkages, 47, 73, 141, 175
lobbyists, **72**, 79
logjams, 41, 95
loss, 23, 43, 83–4, **136**, **139**–40, 142–5, 152–3

melting, 134, 136, 142, 148–9
Middle East, 30, 98, 115
migration, 26, 84, 118, 140, 180–1
mitigation, 12–13, 18–19, 24, 49–50, 54–6, 81–2, 92–5, 100–6, 119–20, 126–7, 130–1, 147–9, 159–60, 174–5, 187–8
modelling, 22
Montreal Protocol, 37, 41, 167
moral obligations, 76, 199
moral, 8, 76, 120, 167, 190, 199, 211–14, 217–18
morally, ix, 2, 8, 13, 69, 209, 214
multilateral, 6, 8, 30, 36, 47, 53, 60–1, 64–5, 95, 212–13

national interests, 4–5, 32, 71–2, 114, 159, 162–3
national identity, 177, 255
natural disasters, 12, 82, 174–5, 186
natural resources, 4, 11–12, 22, 34, 51–2, 106, 196, 206–7, 215, 221, 226
negotiations, 7–8, 31, 37, 41–2, 61–3, 70, 93, 95, 97, 99, 103–4, 107, 169–71, 193–4, 212–3

networks, 34, 40, 55, 67, 71, 74–5, 95, 113, 120, 131, 141–2, 165
new technologies, 3, 15, 54–5, 81, 109–11, 125–6, 196, 208–9, 219
New York city, 36, 171
NGOs, 6, 24, 31, 69–70, **75**, 165–70
norms, 7, 10–11, 37, 58, 64–5, 156, 159, 163–4, 201–2, 224
North–South, 218

ocean currents, 16, 92, 135, 139
optimism, 41, 54, 61, 110, 124, 126, 218, 225–6
orderly conduct, 5, 16, 20, 38, 56, 156, 163–6, 172, 184, 219
ozone depleting substances, 8, 52, 174, 194
ozone depletion, 43, 63, 167

Pacific Ocean, 9, 143
peak oil, 113
per capita, 25–6, 87, 104, 145
petroleum, 107, 116, 124
policy communities, 165
policy cycles, 186
policy networks, 40, 71
political actors, 4, 10, 21, 32–3, 42–3, 46–9, 73–4, 158, 189–91, 194–5, 203–4, 219–20
political association, 13, 28, 35, 68–9, 73, 77, 172, 189
political authority, 6, 16, 20, 35–6, 43, 73, 105, 112, 177, 182, 186–7, 193–4, 197, 204, 211, 224
political entities, 36, 44, 155, 177, 179, 185, 192, 199, 212, 225
political equality, 63
political identity/identities, 6, 71, 163, 169, 198, 213
political leadership, x, 1–2, 14–15, 18, 47, 56–7, 61, 102–3, 127, 129, 189, 194, 223
political obstacles, 85, 108, 218
political order, 7, 29, 35–6, 44, 153, 158, 192–3, 211, 226
political responsibilities, 5, 105
political security, 83, 198–9, 221
political solutions, x, 12, 98, 222, 226

political systems, 11, 18, 33, 43–4, 75, 95, 109, 151
political values, 4, 49, 57, **62**, 66, 68, 165, 206, 214
political visions, 3–5, 15, 49–53, 62, 112, 223–4
political will, 4, 37, 41, 110, 170, 179, 186, 191, 218, 222, 226
pollution, ix, 8, 26–7, 38, 53, 93, 130–1, 154, 206, 220
population density, 145
population growth, 9, 22, 26, 82, 88, 114, 211
poverty, 10, 29, 31, 82, 84, 90, 94, 97–8, 117, 173, 221
precautionary principle, 127–8, 220, 226
predictable surprises, 80, 141, 220
predictions, 25, 81, 100, 115, 157, 170, 173–4, 176, 178–9
predictive models, 173
preservation, 126, 137, 159, 186, 203–4, 209
private sector, 222
productive capacities, 36, 179
productive land, 47, 173, 189
projections, 14, 25, 80, 91–2, 137, 165–6, 208
property rights, 29, 207–8
public education, 47, 175

rainfall patterns, **23**, 26, 92, 94–5, 136, 144, 146, 150, 152, 173, 190
recognition, 6, 10, 57–8, 65, 70, 74, 76, 161–2, 174, 177, 185, 187, 198–201, 224–5
regimes, 8, 33, 47, 164, 214
regulation, 52, 57, 76, 78, 92, 98, 127, 130, 135, 137
regulatory mechanisms, 8, 85, 102
representatives, 74, 167, 177
resource management, 172
resource redistribution, 19, 27
resource security, 21, 29, 81, 84, 122, 131, 206
respect, 6, 19, 27, 153, 178, 184, 212, 219
risk, 4, 7, 34, 38, 49, 67, 81, 92, 99, 123, 127, 132, 146–8, 167, 214, 222, 226
rule of law, 54, 155, 157

salinity, **23**, 82, **136**, 138, 152, 206
sanctions, 61, 65, 211
scenarios, 25–7, 41, 47, 118, 175
sceptics, 24, 161, 176, 208
scientific communities, 78, 164–5, 167, 208
scientific expertise, 79, 161, 167
scientific knowledge, 11, 151, 167, 179, 221
Scientific Revolution, 52
sea level increases, 25, 27, 142–3
security dilemma, 79, 179
security challenges, 77, 98, 122, 128
shared interests, 123, 168, 181, 201–2, 221
social challenges, 52, 145, 153, 183
social costs, 19, 222
social organisation, 34, 38, 47, 96, 156
social progress, 20
social policies, 50
solutions, ix–x, 2–3, 12, 26–8, 30, 54–5, 77, 98, 103, 108, 156–7, 165, 208–9, 211–12, 221–2, 226
sovereign authority, 32, 34, 46, 65, **75**, 78, 104, 106, 157, 182, 189, 225
sovereign states, 5, 18, 29–30, 34, 41, 48, 61, 67, **75**, 78, 161–2, 172, 179–81, 184, 186–7, 200, 212, 224–5
species, ix, 2, **23**, 43, 51, 83–4, 94, 135–7, 140, **144**, 190, 207–9
state sovereignty, 6, 27–9, 39, 177, 182, 188, 203, 219
storm surges, 82, 136, 141, 153
superiority, 44, 208, 216–17
sustainability, 26–7, 81, 109, 121–3, 152, 180, 205
sustainable lifestyles, 111
sustainable practices, 210
sustainable societies, 38

technological advances, 114, 161
territorial boundaries, 29, 50, 182, 211

territorial security, 39, 67, 106, 182, 185, 199–**200**, 211
tidal surges, 92, 143
timeliness, 77, 124, 220
tipping points, 82, 206
trade, 11, 16, 19, 37, **56**–7, 59–60, 71–**2**, 115, 140, 177, 180–2, 195, 206, 219
transboundary, viii, 8, 27, 38, 90, 93, 130–1, 208
transformative, 68, 223, 225
treaties, 7–8, **42**, 164, 181, 199–200, 212, 214
trust, 2, 37, 41, 58, 123, 163, 167, 171, 183

UN, 6, 29, 71–**2**, 84, 87, 129, 164, 166–7, 171
uncertainties, 4, 29–30, 35, 37–9, 49, 114, 128, 161, 168
unilateral action, 36, 191
United States, 103, 121

urban centres, 19, 87, 97, 132
urbanised communities, 141
urgent action, 67, 134

verification, 167
vulnerability, 9, 11, 81, 121, 141, 145, 175, 219

wait-and-see, 15, 102, 128, 182, 220
war(s), 17, 79, 160–1
water availability, 23, 88, 138, 154
water policies, 83, 89
water resources, 16, 28, 32, 45, 83, 87–90, 92–3, 134–5, 137–8, 149, 189
water supplies, 4, 8, 18, 54, 82, 91, 137, 140–1, 146–50, 152
weak drivers, 206
weather patterns, 7, 9, 11, 17–18, 23, 33, 36, 66, 135–6, 142
Western societies, 52, 206–7

Prrinted and bound in Great Britain by
CPI Group (UK) Ltd, Croydon, CR0 4YY

The manufacturer's authorised representative in the EU is Springer Nature Customer Service Centre GmbH, Europaplatz 3, 69115 Heidelberg, Germany. If you have any concerns regarding our products, please contact ProductSafety@springernature.com

Printed and bound by CPI Group (UK) Ltd, Croydon, CR0 4YY
23/03/2026
02076449-0010